Climate Change and Conservation of Coastal Built Heritage

Maya Hassan · Hui Xie

Climate Change and Conservation of Coastal Built Heritage

Springer

Maya Hassan
Faculty of Architecture and Urban Planning
Chongqing University
Chongqing, China

Faculty of Architecture
Tishreen University
Lattakia, Syria

Hui Xie
Faculty of Architecture and Urban Planning
Chongqing University
Chongqing, China

ISBN 978-981-13-8671-8 ISBN 978-981-13-8672-5 (eBook)
https://doi.org/10.1007/978-981-13-8672-5

This Springer imprint is published by the registered company Springer Nature Singapore Pte Ltd.
The registered company address is: 152 Beach Road, #21-01/04 Gateway East, Singapore 189721, Singapore

Preface

The coasts of seas and river banks are the first environment for the emergence of human civilizations, due to the natural and climatic factors. At the same time, they are considered as the first threatened areas of climate changes, which rapidly happen since the end of the twentieth century and the beginning of the twenty-first century, represented by the global warming and the problem of rising water levels. These climate changes will result in catastrophic human consequences and will threaten archaeological sites and cultural inventory in many countries of the world if rapid action is not taken to address them. Conservation of coastal cultural heritage for the next generation and the sustainable and resilience concerning are required. This book aims to develop a protection strategy for coastal areas and built heritage by using both Top-Down and Bottom-Up Processing to face the problems and down to have whole image perception and find a solution.

The first part presented the conservation of coastal built heritage in the era of climate change. Chapter 1 started with the modern conservation principles in the world, China and Syria, then with the shifts of conservation paradigm of the twenty-first century due to climate change and the international crisis of energy. It highlighted the climate change and its direct and indirect impact on the built heritage, and the conservation action levels. The levels hierarchically concatenated from the global action and going down to the national, local, and urban measures, the individual building measures, down to the people power as an effective, fastest, and cheapest action in the preservation process. It emphasized the significance of the participatory among the action levels. Chapter 2 highlighted the monitoring and heritage management methods which started with the advanced leadership and enhancement approaches hand by hand with the documenting, diagnosis, and problems and value identification and assessment. The evaluation approaches and standards for individual heritage structures, the thermal comfort concepts, and requirements have been summarized.

In the field work, the second part of this book followed sequent steps for monitoring built heritage, historical building microclimate, and people satisfaction in two threatened coastal heritage sites from the effects of the climate change in Syria. Chapter 3 tried to develop a protection strategy of the coastal archaeological areas by creating sustainable defensive lines are viable for growth and expansion according to multiple stages of time that commensurate with the increasing problem of rising sea levels. The strategy is based on the exploitation of natural and geographically elements of these areas, with a new idea for economic investment and heritage promotion to have the sustainability of investment defenses step by step whenever widened. Chapter 4 investigated the thermal environment of two typical defensive stone heritage buildings that have been reused as museums in Arwad Island through interviews and a series of field measurement of the summer hygrothermal parameters. Chapter 5 introduced a comprehensive survey of the historic buildings of the old city of Tartous using two methodologies of collecting data. A microclimate study using field measurements of two public buildings, three houses, one multifunction gallery, and one shop, additionally, an indoor thermal questionnaire for the local people and the selected buildings' occupants in conjunction with the field measurements were used. In the end of this book, the conservation strategies and intervention levels were introduced based on Zoom Out–Zoom In approach. Moreover, two conservation frameworks and processes in both macrolevel and microlevel for preserving the threatened coastal historic sites buildings were presented, considering the comprehensive documenting, and environmental condition and problem definition based on research assessments. The DMADV approach in monitoring historic buildings was also introduced to ensure preserving quality and process. The conservation strategies could be generalized in any coastal region in the world is threatened of climate change problem.

Strategic regional planning, field measurement, people thermal satisfaction questionnaire, and data finding based on scientific research and grounded theories are essential tools to have knowledge of the risk size. They form together with a step for assessing the cultural heritage risks and to put outline and guideline for maintenance, restoration and operation, determine problems, difficulties and define where we are now, then shed light on what we need to do in straightway very soon, or on works next years. This book can be used as a reference for researchers, developers, architects, and conservators in protecting the architectural heritage in the coastal areas. It can also be used as the guide for preserving and monitoring the process of the built heritage from both macro- and microlevels.

Keywords Climate Change · Coastal heritage · Conservation Strategies · Microclimate · Thermal Comfort · Occupant satisfaction · Sea-level rise (SLR)

Chongqing, China/Lattakia, Syria Maya Hassan
Chongqing, China Hui Xie

Acknowledgements

This book is supported by the National Natural Science Foundation of China, China Postdoctoral Science Foundation, the Chinese Government Scholarship (CSC), and the Faculty of Architecture and Urban Planning at Chongqing University (CQU). This book would not have been possible without the help and support of many people and bodies. Therefore, we would like to take the opportunity to express our thanks and gratitude to all of them. The Syrian Ministry of Higher Education, Tishreen University, and the Faculty of Architecture in Syria are very grateful. The Antiquities and Museums Department in Tartous and its director and employees, Tartous City Council and its Technical office of the old city and their administration and employees, and Arwad City Council in Syria are acknowledged for their facilities, information, and supports. The authors wish to thank the Coast Guard, the director and employees of Arwad citadel and tower for their cooperation and help. All the locals who participated in our surveys and all the people who allowed us to enter their homes and use their time and properties are gratefully acknowledged. Many thanks to Dr. Tarek Rahmoun for his cooperation and support in the regional planning studies. We also would like to thank Arch. Mirna Hassan and Mr. Mohammad Mahfoud for their cooperation and help in the field surveys. The authors owe everlasting gratitude thanks to their families, children, and CQU Acoustics Laboratory members who have always been a constant source of support, endless love, inspiration, and motivation that we needed to finish our book.

February, 2019

Maya Hassan
Hui Xie

Contents

Acronyms

3ENCULT	"Efficient Energy for EU Cultural Heritage" project of the European Commission
AC	Air conditioner
AHC	Australian Heritage Commission
ASHRAE	The American Society of Heating, Refrigerating, and Air-Conditioning Engineers
Av	Average
BIPV	Building Integrated Photovoltaic systems
BIST	Building Integrated Solar Thermal
BREEAM	Building Research Establishment's Environmental Assessment Method
C	The coastal area
CVI	Coastal Vulnerability Index
DGAM	The Directorate General of Antiquities and Museums, Damascus, Syria
DMADV	Management steps: Define, Measure, Analyze, Design, and Verify
DMAIC	Management steps: Define, Measure, Analyze, Improve, and Control
DPV	The indoor draft perception vote
ECO-Land	Ecological land
EnerPHit	The Energy Retrofit with Passive House Components introduced by (PHI)
EnEV	German energy regulations
EPBD	Energy Performance of Buildings Directive
GCI	Getty conservation institute
HPV	The Indoor Humidity Perception Vote
HVAC	Heating, ventilation, and air conditioning
IPCC	Intergovernmental Panel on Climate Change
LEED	Leadership in Energy and Environmental Design
MAM project	Municipal Administration Modernization project in Syria

Max	Maximum
Min	Minimum
NBA	National Board of Antiquities
NOAA	National Oceanic and Atmospheric Administration
OP	The operative temperature
OUV	Outstanding Universal Value
PDCA	Management steps: Plan, Do, Check, and Act
PHI	The German Passive House Institute
PMV	The Predicted Mean Vote
POE	Post-occupancy evaluation
PPD	The Predicted Percentage of Dissatisfaction
REAP	Rotterdam Energy Approach and Planning
RfF	UK national Retrofit for the Future Program
RH	Indoor relative humidity
SACH	States administration of cultural heritage in China
SLR	Sea-level rise
Temp	Indoor air temperature
Tm	The outdoor mean temperature for Tmin and Tmax
Tmax	The maximum outdoor temperature for the warmest month
Tmin	The minimum outdoor temperature for the coldest month
Tn	Thermal neutrality
TSB	The UK Government's Technology Strategy Board
TSV	Thermal sensation votes
UKCIP	UK Climate Impacts Programme
USGBC	Green Building Council in the USA
WHS	Registered World Heritage Site in the UNESCO List
WHL	World Heritage List of UNESCO

Part I
Conservation of Coastal Built Heritage in the Era of Climate Change

Chapter 1
Climatic Challenges and Conservation Action Levels

This chapter presents the international modern conservation principles and approaches. The principles were shipped to the eastern world in the end of the nineteenth century by the western colonic influence. The differences between the Asian and the western approaches in conservation are presented by the identification of the authenticity issues and determining the heritage values. Paradigm shifts in conservation methods of the twenty-first century appeared and prevailed worldwide due to climate change and the international crisis of energy. This chapter also highlights the climate change and its direct and indirect impact on the built heritage and the conservation action levels. The action levels in terms of the climate change, rising water level, and the calls of energy efficiency are hierarchically concatenated from the global action and going down to the national, local, and municipal levels, then urban and site measures, the individual building and building components measures, down to the people action. The increasing threat of our cultural heritage sites urges to find preserving strategies and to invent a methodology to have a resilient and sustainable heritage. The participatory and cooperation among the action levels is very helpful and effective; additionally, the people power and positive passive behavior could be the significant, fastest, and cheapest action in the preservation process.

1.1 The Evolution of the Conservation Principles

1.1.1 The International Conservation Principles and Approaches

The global heritage conservation definitions started in the western world, in Europe, and later in the Americas. The authenticity, original state, and materials were the main issues in the conservation debates. Since the mid-nineteenth century, the fundamental intervention theories of historic conservation of the built heritage were framed in the dualism of the restoration that inspires a new form that did not appear in the past-

M. Hassan and H. Xie, *Climate Change and Conservation of Coastal Built Heritage*, https://doi.org/10.1007/978-981-13-8672-5_1

time, as exemplified in the work of G. G. Scott in UK, E. Viollet-le-Duc in France, and K. F. Schinkel in Germany. Against this trend, the retention of the status of the modern conservation movement rose which was headed by John Ruskin and William Morris and then sustained by C. Boito and G. Giovannoni in Italy and A. Regiel in Austria in the late nineteenth century and early twentieth centuries. Because of the authenticity's devastation of the historic buildings, the anti-restoration movement criticized this action and worked on preservation and maintenance. Ruskin identified the importance and the value of historical buildings and hence provided a foundation for modern conservation trend. Ruskin absolutely defended the material truth of historic. Sacrifice, truth, power, beauty, life, memory, and obedience were his seven moral lamps of the architecture (Ruskin 1885). He drew attention to the false pride in the new development in urban areas and was worried about the identity losing of the old town. He mentioned the value of the old districts and cities resulted by the ensemble of buildings, space, and different types of details and values and did not depend on only single monuments. In 1904, Madrid Conference set the initial principles of international conservation and emphasized the need for conservation efforts in each country to reach joint work in this field. The conference classified the monuments into two groups: the dead monuments of the previous civilizations and the living monuments which are still in use and recommended the minimal intervention, unity, and stylistic restoration (Locke 1904). The principles of conservative instead of stylistic restoration gained international support. These were introduced into the declaration of the first international meeting on architectural heritage in Athens in 1931. As well, one of the key points of the Venice Charter 1964 determined that the conservation should be based on the authenticity and integrity, and the restoration process should be based on the original material and documents (Charter 1931, 1964).

During the twentieth century, and mainly since the World War II, the cultural heritage conservation has grown as a global goal, covering organizations such as the United Nations Educational, Scientific and Cultural Organization (UNESCO), the International Centre for the Study of the Preservation and Restoration of Cultural Property (ICCROM), the International Council of Museums (ICOM), the International Council on Monuments and Sites (ICOMOS), and the International Union for Conservation of Nature (IUCN). The establishment of guidelines, charts, recommendations, and agreements encouraged awareness campaigns and specialized training activities. The doctrines of Athens Charter 1931 and Venice Charter 1964 and the organizations such as ICOMOS gave the global recognition to the traditional method of the managing and conserving heritage. UNESCO enacted the Convention for Protection of the World Cultural and Natural Heritage in 1972 and created the World Heritage List (WHL) to guarantee the effective preservation of the heritage worldwide. Managing and conserving heritage evolved in different parts of the world in Canada and the USA.

Burra Charter 1979 of the Australia ICOMOS promoted the assessment of the significance of the property based on the values-led approach (ICOMOS 1999). The values-led approach supported not only the advantage of focusing on fabric, but also the focus on a set of other values that are essential to the heritage experts and the stakeholders, where the heritage values are not static, and they change over time depending on the social variable factor itself and the aligning with the shifts in other environmental, cultural, and usage values. The values-led approach to heritage conservation of UNESCO is a participatory process for all interest actors of the property. *It also considers the identification and protection of* the outstanding universal value (OUV) *which is the significance that makes a place essential to all humanity. Therefore, the purpose of managing World Heritage properties is to secure the protection or the long-term maintenance of the outstanding universal place* (Rappoport 2015).

The Venice Charter established the guidelines of the intervention levels in both conservation operations and restoration works. This charter is still being implemented in evaluating the UNESCO conservation projects of the World Heritage listed monuments and historic buildings. On the other hand, UNESCO refers to Burra Charter for the management of heritage sites. The principles of the Venice and Burra charters laid down the general basics and procedures can be applied to any heritage site, regardless of the location, culture, or community. These two principles have confirmed the globalization of heritage that has been and continues to be advocated by UNESCO. Since the mid-1980s, several documents for specific types of heritage have been established, such as archaeological heritage, historic towns, historic gardens, museum collections. Later in the mid-1990s, a global trend opposed to heritage conservation globalization began, with the emergence of the Nara Document in 1994 based on regional and local aspects and a different vision of authenticity based on each society characters, culture, and identity (ICOMOS 1994). Several conservation charters have later versions modification in line with modern theories, technologies, and societies. Figure 1.1 shows the basic international charters and documents for the evaluation of conservation and restoration projects.

A movement from the conservation of individual structure toward the historic urban area has gradually happened, in the meanwhile with the rising concern of the intangible heritage and its significance to the built and tangible heritage. A transfer of principles to the eastern world has occurred through the western colonial and then has changed later, after several decades, to adapt to their culture and concepts. After the international crisis of energy and the lack of resources, the evidences of the impact of changing in the climate conditions on the cultural heritage, and with the emergence of the green and sustainability principles and the rating systems, the conservation principles and approaches started to change and to use the available technologies to adapt to this era needs and to reduce the current and future pressure on the cultural and natural heritage. Paradigm shifts in conservation methods of the twenty-first century have happened based on the value and the sizes of the expenses and risk with a maximum intervention and in many cases to give up the heritage assets under the reasons of saving other life resources and the high costs of conservation which exceed the value of the heritage asset itself.

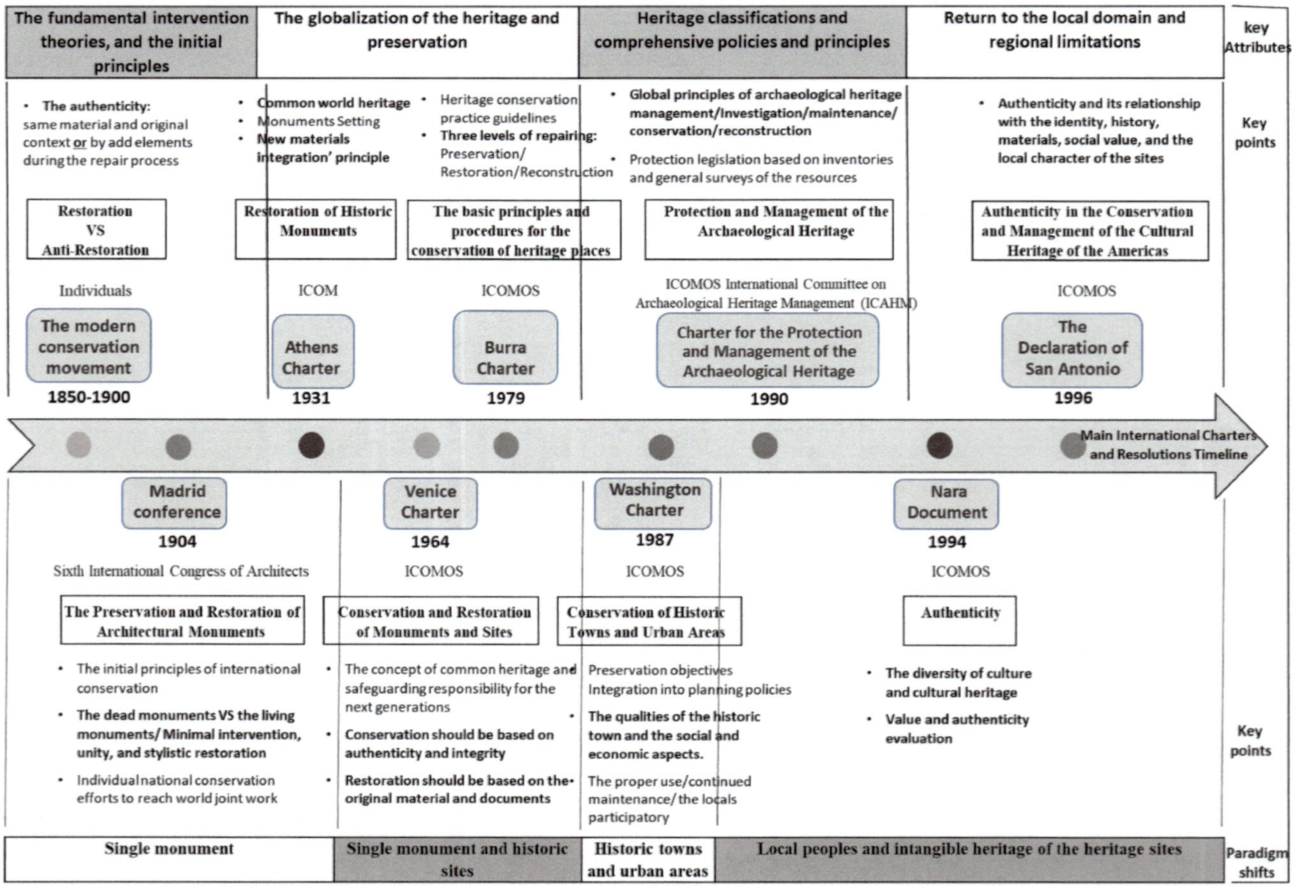

Fig. 1.1 Main international conservation charters and documents

1.1.2 *The Evolution of the Modern Conservation in China*

Modern conservation concepts were shipped to China and the eastern world in the end of the nineteenth century along with the transmission of archaeological knowledge from the west by the western colonic influence. The first law for the protection of ancient objects was promulgated in 1930. During the wartime, from the 1930s till 1940s, as the members of the society for the study of Chinese architecture and the most eminent Chinese intellectuals in the twentieth century, Liang Sicheng, Lin Huiyin, and their colleagues organized the first large-scale investigation of Chinese architecture and initially put forward the modern architectural conservation thoughts according to Chinese context. Liang Sicheng put forward the famous saying: "Don't alter the historic condition" (不改变文物原状), Do not change the original relics, as the modern conservation philosophy of China and later included in Chinese conservation regulations. Between 1930s and 1950s, the initiation of conservation for historic city Beijing and unfortunately the proposal of Liang Chen were not adopted. Since the 1950s, the urban renewal has been gradually employed in the historic center of Beijing, and the city has been expanding dozens of times till n urban problems are getting more and more intensive. The city wall and city gate of Beijing were demolished in 1969. In 1980s–1990s, one of the significant progresses in China's conservation is the definition of built heritage was broadened from single monuments or buildings to historic cities on the national-level legislation thirty years after Liang Sicheng's conservation proposal of Beijing. Authenticity is testified to be valid in China as "not altering the historic condition" is the Chinese interpretation of the doctrine of historic authenticity, Table 1.1. Between 1997 and 2000, China made an international collaboration for heritage preservation such as the State Administration of Cultural Heritage (SACH) in China, the Getty Conservation Institute (GCI), and Australian Heritage Commission (AHC). The promulgation of the first list of national historically and culturally famous cities and the important national landscape and famous scenic zones was in 1982. Since 1985, China imported the international conservation documents and foreign theoretical contributions and documents and started submitting her heritage to the UNESCO World Heritage List. Consequently, till 2010, 40 sites have been inscribed on WHL.

① The Challenges of the Chinese Heritage Last Two Decades

(1) The Chinese heritage suffers from increasing threat from a set of factors related to population growth, environmental degradation, local migration, urban redevelopment, industry agglomeration, and globalization of both traditional sociocultural fabric and the regional economies.
(2) Economic construction and development have destroyed several invaluable heritage sites (Wai-Yin and Shu-Yun 2004).
(3) The environment cannot support enough the human usage that it aims at, causing the deterioration of the structure and destruction of the urban fabric and the loss of the place sense.

Table 1.1 Authenticity in Chinese legislation and scholar

Period	Legislation or scholars	The policy of not altering the historic condition
1930s	Liang Sicheng	Preserve the existing condition or reinstate to its historic condition
1950	The prescription of conservation of monuments and sites was promulgated	Preserve cultural relics in its ancient appearance
1961	Liang Sicheng	"Keep the old as old"
1961	The provisional regulations on protection and administration of cultural relies	Article 11. The principles of restoration to historic condition or preservation of existing condition must be observed Article 12. The principles of not altering the historic condition must be observed
1982	The law of the People's Republic of China on the protection of cultural relics	Article 14. The principles of not altering the historic condition must be observed
2000	The China Principles	Article 2. All conservation measures must observe the principles of not altering the historic condition
2015	Principles for the Conservation of Heritage Sites in China	Article 10 (Authenticity) and Article 14 (Appropriate technology): preserving the historic condition as primary principles using technology with minimum intervention for preventive conservation in the frame of authenticity

Source Guo (2015)

(4) Rapid loss of traditional techniques and knowledge, as well as decreasing of qualified artisans, brings more and more difficulties in maintenance and restoration of monuments and sites. International collaboration introduced modern technology and gained many achievements for the preservation of wall paintings, stone carving, cave temples, etc., but the most widely used repair and restoration for historic timber structures still depend in a large extent on the traditional techniques and artisans.

(5) The excessive tourism and the process of the restoration and presentation for tourism purposes and promotions resulted in new and botched threats to authenticity in China.

② International Debates on Authenticity of Heritage Conservation in Chinese and Eastern Asian Context

There is an ongoing debate about the appropriateness of European approaches to conservation cultural heritage in other areas of the world. The Cultural Charter for Africa (1976), the Burra Charter (1979), and the Nara Document on Authenticity (1994) are notable manifestations of such concerns (Winter 2014). Since eastern Asian architectural heritage mostly is made of timber, bamboo, and other unendurable materials, questions were put forward in the authenticity test of the World Heritage inscription:

How to judge the authenticity after the original materials are replaced? Is material always prior to the test authenticity? The Nara Document is a tacit acknowledgment of the relativities of values and the plurality of approaches to the issue of authenticity of the western notions of intact fabric in respect for the values and the diversity of specific cultural context (ICOMOS 1994). However, it sometimes is misunderstood as tacit of slacking on the conservation of heritage material, even misused by some nationalist, which is possible due to the general nature of this document. Based on the above two situations, two questions were put forward: What China is preserving? And for whom?

(1) The discussion in Nara Conference indicated that the cultural information embedded in cultural heritage should be preserved, and the physical materialized fabric is significant carrier (it is one dimension of the authenticity of that information, but not the unique one). As for architecture built by permanent materials, on the one hand, the original architectural material should be preserved and reused as much as possible in periodical repair. On the other hand, since the replacement of deteriorated material is indispensable, the strict practice of traditional techniques and processes in repair and restoration can assure that the authentic cultural information is transmitted.
(2) The reconstruction of Japanese Ise Shrine, as a specific case related to the living religious tradition, should not be generalized to advocate for reconstruction in heritage conservation. The learned lesson from the case of Ise is that built heritage gains its cultural significance only when it is valued in the associated cultural context; in this condition, tangible and intangible aspects of cultural heritage may not be isolated from each other.

The Nara Declaration of 1994 provided legitimacy to the view that the process of the conservation of heritage is far from universal and depends on the context. Since this strongly influential argument, this view has gained momentum in academic space as well as in the conservation profession, motivating several declarations/subsequent charters; some have implicitly or explicitly asserted the existence of "Asian approach" in preserving the cultural heritage (Winter 2014).

③ Principles for the Conservation of Heritage Sites in China

Between 2000 and 2015, China started to create the self-paradigm in its heritage conservation. In 2000, the China Principles were published based on theoretical and practical Chinese heritage conservation. After fifteen years of continuous conservation works and practice with the increase in the registered sites as provincial, national, and world sites, China showed a high technical level of practice, conservation practice, and a good deal with the different interest actors and stakeholders to use the heritage and achieve the balance between the heritage preserving needs as a non-renewable resource and the other national goals and dimensions. By the end of 2014, China had forty-seven registered World Heritage Sites and extended the concern to the linear World Heritage Sites of the Silk Roads and the Grand Canal and started changing the view to the heritage as power for China development. The Notice on Strengthening the Conservation of Cultural Heritage of the State Council

2005 contributed in organizing annual conferences and forums for sustainable development and usage of the heritage sites and enriching the conservation research. A review of the China Principles was started in 2010, and the contemporary version was established in 2015 with the title "Principles for the Conservation of Heritage Sites in China" (Agnew and Demas 2015). It is a comprehensive system that understands the properties values and practice standards. The new version mentioned the social values and preserving new categories of Chinese cultural heritage such as cultural landscapes, canals, and routes, as well as the last century industrial heritage. It emphasized the conservation of the historic condition as primary principles using technology with minimum intervention for preventive preservation and in the frame of authenticity. Additionally, it formalized the reconstruction of a destroyed historic building for the presentation purpose and the importance of integrated monitoring as a safeguard for the heritage properties and for preventive conservation, where the usage of the properties should be appropriate and adaptive in line with the sites' capacity.

1.1.3 The Evolution of the Archaeology and Heritage Conservation in Syria

① The Development of Archaeology in Syria

The political issues helped in making the archaeological research a priority in Syria and the countries of the Middle East (Gillot 2010). French Mandate provided colonial assistance framework for the evolution of the archaeological research and constitution in Syria based on the historic, political, and aesthetic values. French authorities created the following bodies:

(1) The Islamic Institute of Art and Archaeology in 1918 (Institut d'Art et d'Archéologie islamique).
(2) The Department of Antiquities (Service des Antiquités) and the Standing Archaeological Commission (Mission archéologique permanente) and Damascus Museum (it starts in four rooms in Department of National Knowledge) were also set up in 1919.
(3) The French Institute of Arabic Studies of Damascus established in 1930 (Institut français des Études Arabes de Damas).

Hence, consequently, France benefited from the sharing of antiquities and facilitated the restart of excavations that begun before the World War I, based on new archaeological methods, such as the survey of the Syrian Hills of the Middle Euphrates. From 1919 to 1936, the Syrian antiques were presented in three museums: the Museum of Aleppo in the north, the National Museum of Damascus, and a central Swaida Museum in the south. The French High Commissioner has issued the old Laws of Antiquities in Syria and Lebanon through Resolution No. /207/26 March 1926. The registration on the National Heritage List started in 1929 when they registered Palmyra site as a national heritage. The excavations at Palmyra, Mari, and Ugarit were opened in the interest in Classical Antiquity and Phoenicia under the command/supervision of French officials. Due to the political instability and strategic reasons, the excavation enterprises were risky. In the 1930s, several scientifically programs were developed under notable individuals like Claude Schaeffer (1898–1982), André Parrot (1901–1981), and Maurice Dunand (1898–1987). However, Syrian archaeology continued to be a secondary field of study compared to other countries in the Middle East. In 1938, the first Syrian Law of Antiquities was issued and followed by the issuance of a list of certain milestones, and this list is still yet a reference for historical buildings in Syria. The concept and operations of the restoration and rehabilitation of historic buildings and the public awareness of their importance started strongly to appear.

After the independence in 1946, the consecutive Syrian governments sought to protect/identify a national heritage (athâr) as known by its historic and national or aesthetic value. The modernization/reorganization of the bureaucratic structures that established during the French Mandate was part of a whole process seeking at developing independent national archaeology. The Directorate-General of Antiquities and Museums (DGAM), Damascus, Syria, was founded in 1946. Both of establishment of the Syrian Archaeological Annals in 1950 as a bilingual journal (Les Annales archéologiques syriennes) and the development of the national museums of Aleppo and Damascus had encouraged the development of national archaeology. Several notable national archaeologists participated in the development of the archaeological field in Syria such as Salim Abd al-Haqq, Jaafar Al-Hossni, and Adnan Bounni. The Syrian Antiquities Law and the version governing the work of the staff of the General Directorate of Antiquities and Museums have been issued in 1963. This law was considered by UNESCO as a model law and has been translated into French, and it is still a law in force passed by its provisions to this day and a tool legal for the protection the antiquities in Syria. Later, some amendments had been made in the Syrian Antiquities Law of 1963 in 1969, 1974, and most recently in 1999.

The national archaeology became stronger after Hafez al-Assad in 1970–71, and the protection of antiquities was registered in the Syrian Constitution in 1972. The recognition and hard efforts of the Syrian-related institutions to inscribe the sites on the UNESCO World Heritage List had contributed in enhancing the vital role of international experts in the definition/management of the heritage properties and also in recognizing the universal value of the Syrian national heritage. On August 13, 1975, Syria had signed on the UNESCO's Convention for Protection of the World Cultural and Natural Heritage, while a massive plundering of sites and illegal excavations raised due to a prospering trade in antiquities in Europe and America, which put additional responsibilities on Syrian government authorities to enhance international archaeological collaborations. Between 1980s and 1990s, the archaeological research was characterized by the intensification/diversification, in a political situation that became more appropriate to the polarization of foreign archaeological missions, whose numbers raised to 86 in 1996, while in 2010, there were 120 active archaeological teams in Syria. Some trends seeking the participation of local communities have emerged. The DGAM was aiming to improve archaeology research at the national level within the framework of joint archaeological projects/teams and European training programs alike. This development stage came along with rising Syrian and local concern in heritage protection for tourism development. On the other hand, the monopoly applied by institutions and scientists on archaeological research in parallel with the minimal participation of Syrian civil society and private tourist and cultural groups until the 1990s.

② Monitoring and Managing the Syrian Heritage Sites and Buildings

Since the late nineteenth century, the definition of "heritage" over the world gradually changed from monuments to historic cities in the 1960s, to cultural landscapes recently, thus emphasizing the relationship between territory and man presence. Urban heritage protection dealing initially with the physical conservation of historic cities was extended in the 1980s into functional conservation, with the aim to counteract gentrification, and since 1990s, it is principally focusing on the protection of "place identity" of historic cities, often endangered by mass tourism and the transformation of heritage sites. In Syria, the Syrian Antiquities Law has long concerned the statutory protection of the historic site, the underground archaeology, the heritage assets, and collections and movable objects. Just two decades before the Syrian war, the urban heritage and cultural built heritage started to have its rightful place in the Syrian government concerns and conservation research with attention to their local environment and communities. In 2010, the related authorities proposed a Heritage Law draft and the issuance is stopped since 2011 because of the Syrian war.

Tourism, mainly cultural tourism, has become the third major industry worldwide for a number of employees and contribution to GNP. In fact, it creates benefits in many other sectors. In this context, the UNESCO brand of (the World Heritage) WHS has demonstrated to bring a great added value to the sites assigned and great development of tourism by improving its image and inclusion within networks and plans and the capacity to attract resources. Syria presented the first preliminary list for 16 cultural heritage sites in 1999. In 2011, Syria presented the second preliminary list for 10 cultural sites. Nowadays, six of Syria's archaeological sites were listed on the UNESCO World Heritage List, and other twelve archaeological sites were submitted to UNESCO's tentative list. In 2002, the Budapest Declaration switched the emphasis from the protection of the historic sites to management, and UNESCO invited the listed WHS to prepare management plans. Since 2006, emphasis is placed on monitoring and on the impacts created by the UNESCO World Heritage status, passing from management to monitoring. Therefore, the efforts of conservation changed from Protection to Management then to Monitoring.

Syrian codes outlined the guidelines of the conservation and rehabilitation works without damaging the archaeological structures and resources as follows:

(1) The historic character shall be retained, preserving as much of the original fabric as possible; minimal changes to a historic structure's defining characteristics should be made.

(2) Additions that create a conjectured or false history are not allowed, and the additions throughout the late building's life should be analyzed before decided the necessity of its preservation.
(3) If a new addition is to be made, it is to be differentiated from the old while still keeping with the original structure's architectural features and scale.
(4) The building elements should be repaired rather than replaced. If the repair is not possible, the replacement must be identical in appearance and material to the original.

Recently, the archaeological projects in Syria are done by the foreign missions, the national missions, and the joint missions, which include Syrian and foreign archaeologists and are co-directed by a Syrian/foreign director. Foreign research institutes are also essential participants to coordinate the action/effort of the archaeological missions to enhance international cooperation and also to ensure the continuation of the European and American archaeological research in the Middle East. Despite growing in jointly managed projects, the majority of archaeological scientific production stems from foreign missions, because of the possession of scientific knowledge, techniques, and funds needed to manage exploration and excavation. Syrian society participates in archaeology in two different ways. First of all, the educated and urban elite of Syria, as members of archaeological/heritage associations/societies, they believe that the protection of archaeological remains as a function and responsibility of the citizens and the government alike. As users of the environment of archaeological sites, they work in various activities in and around the archaeological properties as agriculture, archaeological digs, and tourism activity, etc., while some people may participate in the local population as a workforce in excavations and restoration work. These interesting activities are subject to fierce competition among the local population because they form a limited number of jobs and provide seasonal and temporary paid jobs for the locals. Since 2003–2004, a Graduate Center for the restoration of historical monuments and the conservation of natural and archaeological cities opened in collaboration between Damascus University Faculty of Architecture and the French Institute (Chaillot) in Paris, to provide the relative education to the students and of cadres. The Syrian conservation and monitoring concepts are still influenced by the western conservation and knowledge through the collaborated missions, especially French and German. Figure 1.2 shows the evolution of the conservation principles in the world: China and Syria.

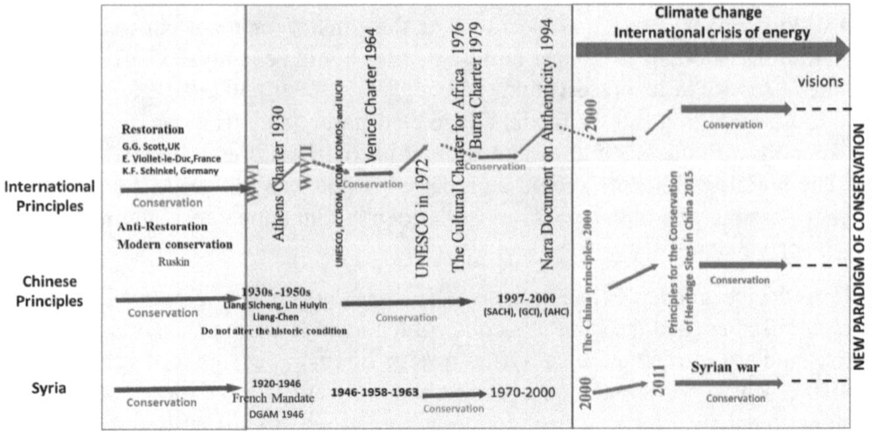

Fig. 1.2 Evolution of the conservation principles in the world: China and Syria

1.2 The Climate Change and Its Impact on the Built Heritage

1.2.1 Global Climate Change and Energy Crisis and the Conservation Paradigm Shifts

The records of the Intergovernmental Panel on Climate Change (IPCC) for the past changes in atmospheric composition over the last millennium demonstrate the increasing of the Earth's surface temperature and the rapid rise in greenhouse gases that are attributable, primarily, to industrial growth since 1750, as well the global average sea-level rise (SLR) between 0.1 and 0.2 m during the twentieth century, It is projected to rise by another 0.09–0.88 m between 1990 and 2100 (Nakicenovic et al. 2000). The average global surface temperature has warmed 0.8 °C in the past century and 0.6 °C in the past three decades (Hansen et al. 2006). According to a number of climate studies, the rate of warming is increasing, and the twentieth century's last two decades were the hottest in 400 years and possibly the warmest for several millennia (Prizzia 2015). Meanwhile, the scientists from IPCC carrying out global warming research have recently predicted that average global temperature could increase between 1.4 and 5.8 °C by the end of the twenty-first century (Solomon et al. 2007). According to the research, the archaeological sites and some buildings have survived at least two periods of global warming around 1500–1200 BC and 800–1200 AD and intervening cold periods. With international scientific evidence mounting and the reliability of future climate predictions increasing (Cassar and Pender 2005), and with the rush urbanization, pollution problems, climate change, and resource consumption, the international awareness about the creation of a strategic solution to minimize the impacts on the communities has increased worldwide.

The first emergence of the sustainable development term in the world was in 1980 in the world conservation strategy of the living resource (IUCN et al. 1980). Then, the famous definition came into the Brundtland Report 1987 "to meet the needs of the present without compromising the ability of future generations to meet their own needs" (United Nations 1987). The sustainable development, with its environmental, economic, and social dimensions and the fourth cultural dimension which added during the Johannesburg Earth Summit of 2002, has been incorporated into national policies of several countries and international organizations. Economic growth caused the threat to the culture and traditions, particularly in the developing countries, and it caused negative consequences on the environment and natural resources. Working separately in each developing sector can achieve specific sectoral objectives, but this may impact gains in other areas, the environmental damage, such as climate change, and can rout developmental benefits. Reducing negative effects, for their own benefit or for saving the planet, is not adequate motivation to influence people's behavior. Behavioral changes come mostly through politics, the economy, or through positive motivation. Therefore, sustainable solutions can be achieved by balancing among the sustainable development dimensions and integrating them (Appendino 2017).

Economic and social developments in the past decades have increased the pressure on resources, including cultural resources, and have placed an unprecedented burden on cultural heritage protectors because of land demand for residential projects or infrastructure projects, a sharp increase in recreational tourism. Old cities which owned a considerable amount of monuments and historic buildings are a core tourist attraction, livable spaces, and identify the citizens' identity, and their values go far beyond their current usage. As sustainable development and green growth have been identified as global priorities in the era of climate change, besides, the need for energy conservation, energy efficiency and the transition to renewable energy. As centuries passed, some historic buildings became uncomfortable and not energy efficient at all. That assures the importance of retrofit the historic building providing the opportunity of getting them back to life with modern versions and facilities. The refurbishment of historic buildings is making a balance between the protection requirements with the need to improve energy efficiency from design to technology and materials. The measurements and monitoring form a basis for the later energy-efficient refurbishment. Finding the weak points is most essential, and monitoring data helps the scientists to develop customized improvements and product solutions and upgrade the historic buildings in an energy-efficient way and in multidisciplinary with technicians and monument conservator. The current tasks to protect the built cultural heritage in the era of energy security and climate change consider:

① Using the measures of energy saving, energy efficiency, and renewable energy to reducing the gas emissions and our impact on climate change,

② Protecting the historic site and buildings from the impact of climate change in terms of increasing intense rainfall, flooding, droughts, heat waves, rising sea level,

③ Providing energy security for the present and future generations in sensible costs, and
④ Adaptation of climate change through the society and several conservation action levels.

1.2.2 The Impact of the Climatic Changes on Preserving the Coastal Built Heritage

Since 1996, one hundred endangered sites of the world monuments watch list need to find protection and adaptation strategies. The proper intervention to preserve the heritage needs a solid base of data collection and accumulated experiences (Pearson and Williams 1996). Due to the challenges that faced both of English Heritage and the National Trust to manage the coastal heritage in the face of sea-level rise, the frequent flooding, and the increase in the expected flood risk in the future of the River Thames on London fabric and its built heritage, English Heritage had the point of the research departure in 2002 to gather evidence on climate change as a possible cause of environmental instability of cultural heritage and to inform present and future planning (Cassar and Pender 2005). On the other hand, the adoption of the UNESCO Convention for the Safeguarding of the Intangible Cultural Heritage in 2003 reflected not only the growing awareness of the interdependence between the world's tangible and intangible heritage and the overall importance of safeguarding cultural diversity; all of these coincided with escalation of international concerns about the climate change risks. In the 29th session of the World Heritage Committee 2005, the World Heritage Centre of UNESCO in cooperation with the Advisory Bodies (ICOMOS, ICCROM, and IUCN) was requested to review the nature and the risk levels that faced the properties of World Heritage because of the climate change. The aims were to propose a mutual strategy to help States Parties to implement suitable management responses and to prepare a joint report on "Predicting and Managing the Effects of Climate Change on World Heritage." The expert meeting was held in 2006 in Paris and resulted in the predicting report and the strategy to protect the world's outstanding values, integrity and authenticity of World Heritage Sites from the harmful impact of climate change. In the 30th session in 2006, the World Heritage Committee took the decision after reviewing these two documents to request all the States Parties to implement that strategy and to develop pilot projects at specific World Heritage Sites, in particular in developing countries to define best practices for the strategy. By combining the four Assessment Reports by the Intergovernmental Panel on Climate Change (IPCC) as the standard work of reference for policy and the predictions from regional climate models maps for future heritage climatologies of the EU Noah's Ark research project and the existing knowledge of deterioration mechanisms, changes in weathering patterns, more frequent flood events, a series of predictive maps of damage, and risk for heritage materials have been determined (McCarthy et al. 2001; Sabbioni and Bonazza 2010; Sabbioni et al. 2010; Cassar 2005). The UK had the first position in the world in establishing the climate change legalization by publishing the UK government Climate Change program in 2006. It also funds a technical unit that supports the work of the IPCC Working Group II on "Impacts, Adaptation and Vulnerability" and the UK Climate Impacts Programme (UKCIP) (Cassar 2009).

The early studies for preserving the natural and cultural resources from the impact of the global climate change focused on collecting the evidences and have a wide image about the problem. In 2007, George Wright Society Conference under the funding of the World Monuments Fund suggested preparing an early triage list of the threatened World Heritage Sites which have the possibility to save and which have not. The threatened sites are classified into three main categories (sites that are doomed, significant sites have to be preserved under any cost, and sites have a possibility to be safe with anticipatory conservation planning). They discussed the need for interdisciplinary cooperation, people's action, and global efforts to develop an integrated approach to effective conserving the built and natural environments (Berenfeld 2008, 2015). In 2007, the UNESCO published a book about 26 natural and cultural sites of the World Heritage List that are threatened by different effects climate change and classified under five categories started with glaciers, marine and terrestrial biodiversity, archaeological sites, and ended with the historic cities and settlements. This book also presented the adaptation measures of some sites (Colette 2007). Several local evidence around the world and the global responses were discussed based on the examining of cultural heritage sites of the 2008 Watch List that are threatened by the impacts of climate change, in sites such as Herschel Island in Canada, Sandviken Bay in Norway, Antarctica, Louisiana, Mauritania, Senegal, Tanzania and Gambia in Africa, Himalayan region in India, and Sonargaon in Bangladesh. Melting ice, drought, desert sand storms, flash flooding, sea-level rise, temperature fluctuations, and soil and coastal erosion affect the built and natural heritage and its communities and their activities. The negative impacts of climate change will have consequences for humanity as a whole including the products of human creativity. Several changes in climate will impact on preserving the archaeological heritage: firstly, the direct physical effects on the built heritage, and secondly, the effects on social structures and habitats migration, and societies which are currently maintaining the sites.

① The Direct Physical Effects on the Built Heritage

(1) Historic structures are characterized by a greater intimacy with the ground than modern ones. The porous materials of their walls and floors cause drawing water from the ground and then losing by surface evaporation causing corrosion or salt weathering as secondary effects (Colette 2007).
(2) The circular of drying and wetting in nature and it's different circumstances affect the crystallization and dissolution of salts, thus affecting buried remains as well as frescoes, murals, other ornate surfaces, and rock art.
(3) Today, historical rainwater systems suffer from multiple problems because of their inability to deal with the amount of heavy rainfall which has increased over time, and often, the main reason is difficult to access, modify, or maintain them.

(4) Several factors such as desertification, erosion, and salt weathering cause damage to the cultural heritage in desert environments.
(5) Seasonal variations of temperature and humidity and the amplitude of the diurnal besides the shocks (extreme and sudden variations) are key factors affecting building materials and surfaces and causing flaking, splitting, and cracking.
(6) Organic building materials, especially timber, will be directly affected.
(7) The problem of sea-level rise as a result of climate change directly affects coastal areas by coastal erosion, increase in the sea salt chlorides for coastal soils, and continual submersion of low areas.
(8) The erosive character of rapid flowing and invariably dirty of the flooding water can harm building materials not designed to withstand prolonged immersion. Additionally, post-flooding drying could motivate the harmful microorganisms' growth such as molds, fungi, and staining.
(9) Coastal erosion and significant retreat of the shoreline will threaten coastal properties with total loss.
(10) Increases in storms and winds will result in structural damage to whole structures and vulnerable elements.

② The Effects on Social Structures and Habitat migration

Climate change is a basic threat that has direct physical effects. These impacts have cultural and societal results on the communities of the cultural sites and on the dynamic heritage of the landscapes and structures where the people work, live, worship, and socialize (Colette 2007):

(1) The disintegration of communities and the property abandonment as a result of climate change, with the loss of cultural and ritual memories in the end, can also force the citizens to migrate under the pressure of flooding, sea-level rise, or desertification.
(2) The abandonment is a great issue in the contexts, where traditional skills and knowledge are necessary to guarantee the proper preserving and maintaining of the cultural properties.
(3) Biological changes with species shifting ranges have also an effect on conservation operation with limiting the availability of native species to repair buildings. There is a need for a strategic shift from a reactive approach to a preventive and proactive approach.

③ The Risks of the Global Warming Problem on the Coasts

The problem of global warming carries two main risks: The first risk is the environmental risk for increasing drought-prone areas and the melting of the Arctic ice. This will lead to SLR, and the second risk is the demographic risk because of population movement beside bodies of water, such as rivers, seas, and oceans. Because of the high proportion of drought in the interior areas, this will lead to environmental migration, while international reports indicate that it will probably constitute a coastal population, half the world's population in 2025 (Creel 2003; Costanza et al.

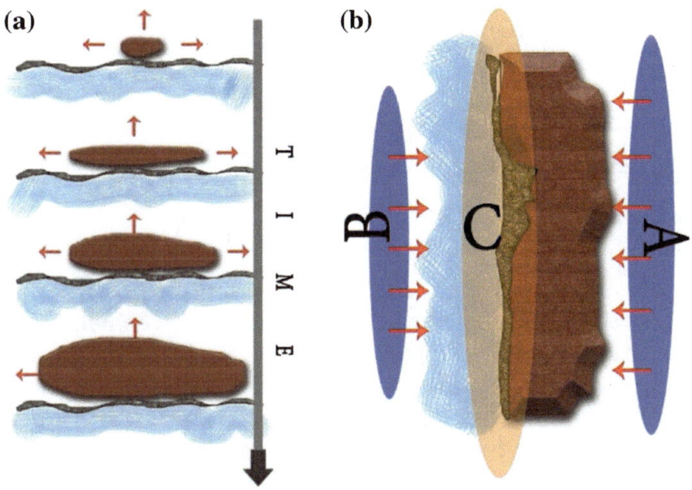

Fig. 1.3 Risks of the global warming on the coasts: **a** the growth of coastal cities, the presence of the specified (water body) and **b** the challenges in the coastal areas

2011). Due to these risks, coastal areas and coastal cities will face two types of challenges:

A. **A surge of coastal city growth**: Besides having innate qualities in the emergence of the coastal cities (good soil for agriculture, mild climate, trade, freshwater springs), there is additional external pressure on coastal cities (climate change, rising rates of drought in the inland areas, population growth), as shown in Fig. 1.3a. Ultimately, this large population pressure will lead in accordance with Global Reporting to the rapid drain of natural energy sources, along the coastal areas (Rahman and Rahman 2015).
B. **On the other hand, global warming can lead to the increase of natural disasters**, starting from floods and hurricanes in addition to the big disaster of SLR caused by the melting of continental ice, which is expected to lead to the disappearance of some islands and parts of coastal cities along the world's coasts by the year 2060 (New et al. 2011). SLR could affect millions of people unless the states work together to reduce carbon emissions seriously.

The challenges (A and B) as mentioned above could be solved with strategic plans for the coastal area (C). Different solution strategies are directed to the specific coastal characteristics and problems. This may encourage new development for urban growth in accordance with environmental rules, as shown in Fig. 1.3.b. Consequently, the proposed defense strategy can be applied to face the SLR problem within the area (C). Promoting the continental shelf possibilities will lead in the end to protect the existing coastal areas and contain expected human activities in the future within the areas of progress to protect the threatened areas. The strategies will ultimately help with the emergence of new axes development for the coastal region as a whole (Rahmoun et al. 2016).

1.3 Conservation Action Levels for Coastal Built Heritage Against Climate Change

Adaptation of climate change consists of actions through society, by individuals, groups, and governments. These levels of operations run within hierarchical structures, where the levels interact with each other (Smit et al. 2000; Adgera et al. 2005). The hierarchical structure extends beyond the nation-state to the international scale (Protocol 1997), Fig. 1.4. For that reason, the success relies on the scale of application and the criteria used to assess it at each level.

National and regional strategies, policies, and measures to minimize the climate hazard have to set to enhance the future security such as the technical effectiveness of flood protection defenses as a component of long-term sustainability for the public good. The short-term planning is based on self-reliance and preservation of local identities, the concentrating on the economic efficiency in adaptation action, and welfare maximization at the aggregate-level emphasis on nationally agreed objectives for organizations or individuals. Finding the balance between levels and developing a multi-level/cross-scalar approach for coastal zone's management and by encouraging new forms of collective participation in decision-making processes are the ways for success, where the time plays a significant role in the effective implementation. The climate change adaptation should be observed in markets within resource-based

Fig. 1.4 Conservation action levels for the coastal built heritage against climate change

livelihoods and other levels such as water resources, coastal planning, health interventions, and built environment (Adgera et al. 2005).

1.3.1 Global Action Level

The international challenges have forced the international and national bodies to involve their participatory and legalizations to adopting modern tools of problem diagnose and process improvement. The rash technical progress of the last years and the always increasing awareness of energy saving and communities identities and heritage made the road to improve our heritage is becoming almost mandatory to be able to survive the always increased costs of raw materials and energy (de Almeida 2014). Numerous summit meetings, conferences, treaties, and conventions have been held to reduce the impacts of climate change. Additionally, the World Bank and international associations for data resources, such as the National Oceanic and Atmospheric Administration (NOAA), the Intergovernmental Panel on Climate Change (IPCC), and NASA's Earth Observing Satellite System, opened their reliable data resources to all the people and researchers to encourage the contributions worldwide in the aim to minimize the risks and gather the efforts to find the proper adaptation. A short review of the notable global efforts toward the international and significant global warming, rising water level, and energy consumption is presented in the following paragraphs.

① **Global Scenarios for SLR Problem**

(1) **Adjustment Strategy**

Adjustment strategy has protection measures for the coastal areas from flooding and SLR risks by encouraging the development and adaptation and adjusting policies for the existing threatened buildings to be more resilient. Ancient adaptation strategy is used for temporary leaving of the early human settlement on the threatened riverbanks and coastal areas during catastrophic flood events. It was planned by the locals themselves based on their needs without any policies intervention, and then, it was changed to be planned in advance. Currently, the measures of this strategy are more developed and flexible and could combine measures of later strategies to have continuous development (Bray et al. 1997; Horowitz 2016). Using this strategy minimizes the action costs based on the improved technologies and warning systems, as well as the simulation outcomes which determine the safe area and water movement in extreme weather (Kourgialas and Karatzas 2014). It also has economic development and protects the existing accommodation and local communities (Roggema 2009; Horowitz 2016). The implementation of this strategy is based on building codes, organizational policies, raising buildings and land to the minimum level of construction based on the predicted rising water levels, and the pre-identified high flood risk zones (Bray et al. 1997; Few et al. 2007), where the predicted threatened structure should follow the restrictions of using the first floor or raising the building by redesigning the first floor such as an empty space and in some cases moving all the valuable furniture and personal things from first floor to other place and continue using it in safe time and then move to the upper floor or other place based on the warning systems to minimize the risk during flood and storm events (Grannis 2011; Wamsler et al. 2013).

(2) **Defense Strategy**

The defense strategy has protection measures to protect the coastal areas from the floodwater and provide preventive preservation using hard or soft measures and policies. Building the coastal flood shelters on stilts and early warning systems are the early sustainable flood mitigation methods. Hard measures include permanent structures, embankments, dykes, breakwaters, construction of tidal walls, fixed water land, and barriers or movable barriers and floodgates (Bray et al. 1997; Klein et al. 1998; Kim et al. 2012; Lee 2014; Sköld et al. 2015; Horowitz 2016). Although this kind of measures is so efficient, increases the land area, and minimizes the erosion of the protected coast, marine ecosystem should take into account during the planning process to minimize the aesthetic and negative ecological effects and water moving in normal and safe time (Bray et al. 1997; Ford and Ford 2011; Lee 2014; Horowitz 2013, 2016). On the other hand, soft measures include temporal or permanent infrastructure and low-cost flood barriers such as berms, excess water absorbing measures, wetland reclamation, and beach nourishment (Bray et al. 1997; Grannis 2011; Nicholls 2011; Horowitz 2016). Floodwater could be pumped out of the interior of low lands on a daily basis by pumping stations. This soft defense strategy is more eco-friendly and has less coast than the hard one (Lee 2014). The hard measures have several early implementations and policies around the world such as the Oosterscheldekering movable barrier in the Netherlands with three movable flood barriers and two artificial islands: the Thames Barrier which protects London Heritage which is estimated that it will need to be closed about 200 times per year to protect London from tidal flooding by 2100, the MOSE flood movable barrier to protect Venice from sinking, the Fort Massachusetts in the National Park Service in USA, and restoring the natural barriers of Gereza Fort at Kilwa in Tanzania (Colette 2007; Berenfeld 2008). The implementation of this kind of strategy should have an interaction with the country planning levels after a proper analysis of the location land use and cover (Bray et al. 1997).

(3) **Retreat Strategy**

Retreat strategy has abandonment measures in high flood risk coasts by moving communities and economic and valuable structures from the places under the risk to safer places. This strategy tries to adapt the climate conditions when the flood hazard and vulnerability are such high, where no other effective measures of survival (Lee 2014; Grannis 2011; Nicholls 2011). The implementation should be planned by the government in advance in cooperation with the locals (Kershner 2010; Bray et al. 1997). Many examples of this adaptation around the world are moving the heritage structures in the Canadian Herschel Island Territorial Park due to the progression of the shoreline, the American Cape Hatteras Lighthouse on national park, and the famous moving of temple of Abu Simbel to save it from flooding in Egypt. The risk scale and the impacts of climate change linked to the time should identify before the moving action implementation. Not all the cultural resources could be moved due to moving obstacles/problems of huge and ancient structures, losing authenticity, high cost, and practical reasons.

(4) **Attack Strategy**

Attack strategy has measures to increase the land of the coastline by developing several water projects and build infrastructures and bodies inside the water, where people can move to live in there in order to reduce the coastal area population pressure by redistributing its population on the new structures and lands and save the people from the flood risk. Several advanced techniques could be used to protect the new bodies and structures from flooding risks (Building Futures & ICE 2010; Graeme 2015), such as the amphibious and floating structures with floating techniques, foundations, or platforms, or movable foundations can go up and down according to the water level to protect them from flood risk and prevent them from sinking (Hendriks et al. 1999; Rijcken 2003). New trends and ideas appeared of the floating communities at the beginning of this century, but most of them are still on the papers. The Netherlands historically suffered from the flood risk and all the world learn the lessons from Dutch people and their experiences. The Dutch project of the floating city IJmeer on the Rhine Delta formed a comprehensive project for planning and designing a protected safe city with all the housing facilities and floating vegetation and increase in land area with a description of the implementation policies and different scales (Graaf et al. 2006). Some preservation projects and protection scenarios of the impact of climate change have high costs, that why Berenfeld mentioned the triage strategy for saving the cultural heritage for emergency planning with allocated efficiently funds (Berenfeld 2015).

② **Global Action for Energy Efficiency**

In the EU-27, 14% of structures were constructed before 1919 and 26% before 1945. Some of them are listed structures due to the historical significance and protected by conservation policies. Buildings are considered as one of the main energy consumers in the urban context. The energy demand for buildings in Europe is as high as 40% of the total energy demand. Finding energy-efficient conservation solutions to renovate historic buildings and centers of the cities promotes the long-term preservation and sustainable management of our cities and provides climate protection and indoor comfort living conditions (Troi 2011). Creating buildings equipped with integrated sustainable energy solutions is the key to achieve more energy-efficient and environmentally friendly buildings (Tomás et al. 2010). Energy legislation is an assured tool to aid in reaching energy ambitions. The first version of the Energy Performance of Buildings Directive (EPBD) was established in 2002 at a European level and affecting the regulations in all Member States in regard to energy consumption in the built environment. The EPBD imposes a required evaluation of energy-saving possibilities in the historic structures. The concept behind such a required analysis is to have at least an early stage of energy-saving actions during the restoration procedure of the

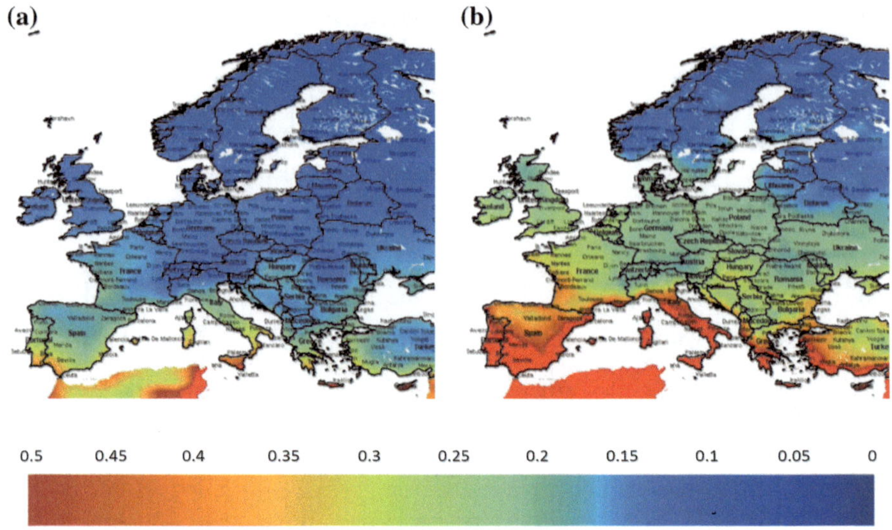

Fig. 1.5 Cost optimal component quality (U-value) **a** for basement ceiling insulation for a new building and **b** for a historic building in which only the basement ceiling can be insulated. *Source* Bastian et al. (2014)

built heritage and support the innovations of energy-saving technologies and products for historic structures. The EPBD legislation came out in conjunction with the conservators' worry about its impact in disfiguring or ruining the historic buildings. Currently, this measure is worldwide understood and needed to make the structures more energy efficient in a way that is compatible with the heritage value of a building (Bastian et al. 2014), Fig. 1.5. Some European countries imposed an energy certificate for listed historic buildings, and the suggested measures provide a logical possibility to replace some components of the structure with high-efficiency items. The listed historic structures under specific conditions could be excused from national minimum energy performance requirements for the components of the building envelope that are retrofitted or changed.

The climate and location have an essential influence on the building performance. The passive design and strategies minimize the annual thermal load of the structure (Gong et al. 2012). Based on optimum analyzing of a total of twenty-five representative cities, a total of seven climatic zones in China were mapped, and seven passive design measures for each zone were optimized as a guide for architectural design to reduce the energy consumption: wall thickness, external wall insulation thickness, window orientation, window wall ratio, glazing type, sunroom depth/overhang depth, and roof insulation thickness (Gong et al. 2012), Fig. 1.6. European climate zones have been mapped and classified, and a set of requirements have been established to achieve a bioclimatic design for each zone (PVSITES 2016). Koppen-Geiger world map of the climate classification and its updated maps in addition to the national

Fig. 1.6 Chinese passive design zones based on optimizing of thermal load performance. *Source* Gong et al. (2012)

climate classification maps could help to establish the regional map and the overall thermal legalization (Peel et al. 2007).

1.3.2 National, Local, and Site Action Levels

① National Action Level

Several European local preservation bodies established national guidelines and tight restrictions in regard to the energy demand for the new structures and throughout the restorations. They consider district heating by combined heat and power plants as a proper solution for saving energy because of its little effect on the building, and the solar panels are not the good choice if they destroy the entire roofscape of old cities. The UK Climate Impacts Programme (UKCIP) for the climate change scenarios and research is considered one of the world's leading programs. The Energy and Climate Change Department was founded to implement the government's policy in 2008. The UK Climate Change Bill made UK the first country in the world to devote in law the goals of climate change and contributed in making it a trustable country in the reputation for advice-giving in brand change systems, energy labels (Cassar 2009). The UK Government's Technology Strategy Board (TSB) funded the national Retrofit for the Future (RfF) program from 2009 to 2013, to find innovative renovation approach of social housing stock, utilizing a "whole-house" method for achieving an 80% CO_2 emission decrease target for 2050 to reach footprint close to zero for all the structures. A vital request to achieve these targets is to understand the potential and effectiveness of retrofitting the different housing types (HM Government 2011; Palmer and Cooper 2013). Besides the English heritage and UK legalizations, many cities in the old continent are discussing how to optimize energy efficiency moving away from fossil fuels and how to implement technologies and stimulate behavior to achieve the sustainable energy in their cities' master plans. Many learned lesson could be gotten from the European experience in the energy efficiency field and the retrofit of the existing and historic buildings in the same context, such as the German Energy Retrofit with Passive House, the 3ENCULT "Efficient Energy for EU Cultural Heritage" project of the European Commission in both building technology and local governments' levels, as well as, the Netherlands experience in rising sea-level action and the use of renewable energy.

The German Passive House Institute (PHI) introduced the Energy Retrofit with Passive House Components (EnerPHit) standard in 2010. The fundamental concept is to improve all relevant elements of the old structure with Typical Passive House components through the retrofit process to reach high energy efficiency with occupant comfort at minimal expenses. The EnerPHit voluntary standard for advanced energy retrofit might just be related to retrofits of non-listed structures in cool, temperate climates like Central Europe. It includes an efficient heat recovery ventilation system, windows with triple glazing and insulated frames, more than 200 mm of thermal insulation and very good airtightness.

The 3ENCULT project developed the EnerPHit standard to also allow for the listed and historic structures certification. The development contains special provisions for historic buildings with valuable façades, where the renovation can only be with interior insulation. Additionally, exemptions for other parts of the building with restrictions by the cultural heritage authorities were introduced. The concept aims to improve the efficiency of each part of a historic structure as long as this corresponds to the protection of the cultural heritage value. Further development under the 3ENCULT is the adaptation for application in all European climates. The quality requirements for individual building components will remain the foundation of the EnerPHit's international standards. The economic improvement process was the leader for the new component requirements in different climates. The process was performed for each location in a network of climate datasets covering all parts of Europe, with the aim of finding a set of component qualities with the lowest life-cycle costs for an example building. A total of 200 combinations of different ventilations, windows, and shading qualities were combined with different insulation levels of the opaque building envelope. Thus, a set of components can be defined that leads to a lower amount of investment and energy costs using the net present value method. The 3ENCULT project attempted to bridge the gap between preserving historic buildings and dealing with climate protection by assessing the needs of historic buildings, and then, proper strategy or technical solutions were selected for delivering the building energy retrofit based on monitoring and field measurements in eight case studies in different European countries and climatic zones and from different historic eras. The project founded a technical guide on energy-efficient renovation of European historic structures and the guidelines and recommendations on integrating energy-efficient retrofit of historic buildings into policy and planning of the local governments at the municipal level. The main technical solutions in the building level were on its envelope, windows, ventilation, and passive and active energy-efficient solutions (Efficiency Energy for EU cultural heritage 2013). In summary, the countries have to face individually and unitedly the common issues to inform future retrofit efforts and climate change problems.

② The Local Government and Municipality Action Level

The local government and municipality have a significant and central role to be the developer of local strategy and policy address the energy-efficient historic buildings and districts renovation in a comprehensive approach toward sustainable development at the local community level. It can plan the sustainable energy transition in the community, reduce energy demand and switch to renewable energy, and encourage the local community to engage in the sustainable energy transmission. The municipality might support the adaptive reuse, rehabilitation, and retrofit projects for historic buildings, listed and certified properties by establishing local credits. The local credits must give the owner more cost-effective option than rebuilding and provide him additional savings and cost advantages of rehabilitation projects. They differ depending on their location and a number of provisions to be eligible for such a credit (Brandoni and Polonara 2012). The municipality staff and citizens by acting together can make the difference between failure and success. Where the project dominated by local government, the municipality and civilian initiative both can be successful. Hybrid cooperations of both are less successful (Sanders and Timmeren 2016). Only urban planning based on the sustainable rules can generate an effective sustainable architecture (Gauzin-Müller 2002).

Municipal strategies improve the resilience through mandatory retrofitting of the building and mandatory building codes for new construction and main renovations, the motivations for voluntary work, funding mechanisms, and grants to simplify voluntary action, education, and outreach efforts (Linnean Solutions, The Built Environment Coalition, The Resilient Design Institute 2013). Several lessons can be learned from the Dutch municipal mapping initiatives in cooperation with the research bodies, such as Rotterdam city scan and the energy potential mapping of the Northern Netherlands, which form the base for the proposed interventions of the city map. Mapping the energy potentials helps sustainable spatial planning based on the effective use of local energy capabilities, particularly in developing countries (Broersma et al. 2013). Additionally, the Rotterdam Energy Approach and Planning (REAP) used swarm planning as another nature-based principle for rapid changes toward sustainability. The REAP considered the city fabric as an organism that has an ability to intelligently exchange energy resources and waste flows. It focuses on cities more than buildings, where the individual structure can have independent work alone and fully autonomously resources work in the proper sustainable urban context (Dobbelsteen et al. 2010).

③ **Urban and Site Action level**

The vernacular urbanism and architecture, as well as the courtyard houses worldwide, used water and traditional vegetation techniques to adapt the local climate and provide the occupant comfort. Recently, the recommendations of LEED and BREEAM have demonstrated that the possibility of the sustainable refurbishment of heritage buildings begins with developing the site and overcoming the potential barriers and renovation challenges (Sharifi and Murayama 2013; Balson et al. 2014). The microclimate and urban local climate and comfort are influenced by several essential factors such as the urban form and landscaping type, site buildings, surfaces, vegetation, soil, and weather. The anthropogenic changes in land cover and built forms caused heat island and urban eco-stress over the world (Arnfield 2003). Increasing the amount of vegetation, plant type, and arrangement and providing cooling through shading and evaporation could help in improving the local climate, reduce the effects of high heat, and enhance the temperatures during the day and night (Guhathakurta and Gober 2010). Middel et al. found that the dense urban forms in the desert can create local cool islands in mid-afternoon in semiarid Phoenix, Arizona (Middel et al. 2014). Other efforts made a proper treatment for the land surface after parameters analyzing for the prevalent soil type for the study sites which have courtyards, yards, or privet garden. The vegetation types such as lawns, perennial and non-perennial trees, shrubs, and vines are used to enhance temperatures in the landscape. A typical summer grass called Bermuda (Cynodon dactylon) with heat-drought-resistant properties and Mulch (decomposing granite) are frequently used for non-grass surfaces in landscaping the desert in the southwestern USA to minimize maintenance and the soil evaporation rates (Middel et al. 2014). The conservation site policies must consider the buildings under danger. American New Orleans and south Louisiana provide flood protection using the Buoyant Foundation Project and amphibious houses near the existing coastline. Where the floating houses go up and down by the foundation which can raise the structures as high as it needs, the occupants use ground floor as car parking and unimportant function the upper floor to live and protect themselves during the flood time (Ovink 2014). Sustainable infrastructure and neighborhood could be the catalyst and guarantor of the process of individual development of the architectural vocabularies.

1.3.3 Building Action Level

Since the UN report by the Brundtland Committee, sustainability in the built environment has primarily focused on the technical side of single products/structures (Dobbelsteen et al. 2010). This section will introduce each of the energy efficiency measures, retrofitting and adaptive reuse, and some solutions to upgrade the historic buildings.

① Energy Efficiency and Historic Structures

The energy efficiency could be easily achieved in the new buildings when all aspects of their design and construction are planned to be sustainable from the outset. The great challenge is how to deal with the existing buildings, and far more critical challenge is how to handle the heritage buildings without affecting their values. According to Borges et al. (2014), there are four strategies to handle with older building stocks: Firstly, buildings can be preserved as they are, accepting in the process that energy consumption will be much higher. The second strategy aims to integrate refurbishment measures into the building in such a way that the original structure can be optically and technically preserved as much as possible. The third strategy for dealing with the existing buildings is the synthetic reconstruction of buildings that can no longer be supported from a cultural–architectural point of view or because of their poor condition. The fourth strategy involves the deliberate separation of old and new but united in an overall composition, where the old and new are kept apart in two distinct buildings in one dialogue. Although the urban architecture and urban heritage conservation are not in the focus of the German energy regulations (EnEV), special regulations for energy in built heritage have been established connected with renewable energies and energy consumption issues, where saving the environment is a basic root of long-term heritage conservation. The monumental and listed buildings and the amount of energy efficiency should be selected with particular respect to the authentic material and construction, and they can make energy consulting instead of energy rating to find compromises (Kaiser 2009). The energy efficiency retrofit of the historic buildings not only provides them the chance of survival but also helps in prolonging their ages and usage through the following aspects:

- The proper maintenance and management of historic properties as livable spaces for living, working, and entertainment,
- Providing the occupant comfort hand in hand with the maintenance of historic aspects,
- Enhance the structure protection against climate change,
- Energy efficiency retrofit could reduce energy needs and costs, and
- Helping in the create eco-friendly action and behavior of the occupants of these buildings and within their urban area.

② Retrofitting and Adaptive Reuse of the Historic Buildings

The heritage structures are constantly advancing in age and outperforming their original functions. Moreover, the needs, lifestyle, and technology change and the historic buildings need to keep pace with changes. The restoration costs and the sensitivity of their structures increase the difficulties of preserving task in order to meet the modern demands. The prediction of the possible risks is important to consider the appropriate actions for preserving through different conservation processes and interventions. Conserving of the cultural heritage can extend the life of these valuable resources of past cultures as well as it can improve the economic conditions of a society. Most cultural heritage assets are non-renewable so the intervention should be able to achieve the contemporary requirements without altering the intrinsic values of them (Magrini and Franco 2016).

Adaptive reuse is a nondestructive process of retrofitting historic structures for new usage concerning their historic integrity and meets the modern needs to breathe life into our heritage and save its continuity. The social, environmental, and cultural advantages of adaptive reuse hand in hand with the sustainable development requirements give an added vibrant value to our built heritage and community. The restoration or conservation process includes restoring a structure to its original state, and it differs from the adaptive reuse which changes the intent of a structure to meet the modern needs. The historic skin value with the structure adaptive reuse revitalizes the community by using the intrinsic cultural value and marketing growth and its profits. However, the cultural dimension, value, aesthetics distinction, location, design, facades' details, and materials of historic buildings are superior to the modern buildings. The budgets and high expenses for the retrofit project are the essential factors in determining the owner reuse decision, where not all structures are ideal for the retrofit. Reuse the historic building in many cases is done with the investment objectives of the decision-makers without much consideration of the buildings character and the end users' needs. Although the economic factor plays a big role in refurbishment and reuse measures, the urgent need is to provide proper occupants' comfort while keeping its historic significance and value.

The measures of enhancing the indoor air quality of the existing structures should rely on the building usage types such as the residential, educational, commercial, and official use as will be shown later in the thermal comfort and indoor air quality standard section. The definition of the new function should be in the preservation context of the building structure and characters and in balance with its nature and its community needs. Economic, cultural, social, and physical factors influence the selection of the new function. The use of the historic buildings as dwellings currently is declining as more and more people decide to move to new houses, which they believe their life would be better and convenient. Understanding the required changes in the building and its cultural significance to meet the needs of the new function shall be done before the restoration work. Reusing historical building in a proper way is very important because it helps to maintain the building physical properties.

③ Common Interventions to Upgrade the Historic Buildings

Most of the current researches recommend the passive strategies and measures to achieve energy efficiency and better comfort indoor environment. The use of solar energy and passive strategies in urban planning and buildings' refurbishment should be based on the opportunities of the location, climatic data, and the buildings themselves in the line with the preservation requirements (Rabah and Tamkan 2002). A study in the historic buildings in Catania in Italy showed that the applying of the energy retrofit interventions of the solar technology systems of the Building Integrated Photovoltaic (BIPV) systems and Building Integrated Solar Thermal (BIST) can reduce the energy consumption of 19% during the winter and of 32% during the summer (Moschella et al. 2013). The traditional building elements such as the existing of wind catcher, courtyard, and basement, and the past lifestyle and behavior of the inhabitants are effective solutions for improving thermal comfort and minimizing the energy usage and can inspire the new design and architecture (Khalili and Amindeldar 2014). Developing windows insulation with very high low air leakage, by dividing the functions into the insulation for the inner layer and the aesthetics for the outer layer, can also contribute on reducing the energy usage and save the historic value (Efficiency Energy for EU cultural heritage 2013).

A study in Egypt showed that the thermal comfort in the Ottoman residential buildings is better than the houses of the nineteenth century and emphasized the importance of having an inspiring solution from the historic buildings to the new designs in Egypt toward a green architecture (Saleh et al. 2017). Another study also showed that the Ottoman residential buildings used building techniques and interior spaces distribution of design respect the environmental conditions and the current green architecture principle, as well as the light-colored materials have high thermal insulation properties, which decrease temperature (Mohamed and Ali 2014). The dense urban fabric plays a significant role in reducing the direct solar radiation and providing a better indoor temperature of the historic residential buildings in old Cairo in summer by approximately nine degrees lower than the residential buildings in a vast urban fabric (Shalkany 2014). The recent trends of conservation have given up the notion to completely preserve historic buildings to develop further. A study in the city center of Seville in Spain tested three historic residential building restoration alternatives: original state historic structure, model restored by intervened building materials, which is in compliance with Seville preservation regulations, and the third model restored by the sustainable and eco-efficient building materials. The third model is distinguished by better performance in winter and cooling in summer, and better reducing CO_2 emissions by more than 50% during the life of the building (Gálvez et al. 2013). The technical treatment could preserve the structural aspects. Each case has to be seen individually without conflicting the historical character. Improving a building's envelope by improving some parts such as the walls or windows is typically one of the challenging ways due to its impact on the appearance, but it develops the building performance. Urban and landscape intervention and sustainable infrastructure that improve the building envelope and lighting methods, insulation for the inner and outer layers, windows upgrading, lime plaster, roof solar panels, water calculator, and the spaces' ventilation to minimize the moisture are the common interventions for upgrading the heritage structures, as well as reducing the reason for water entering to the structure by keeping the roof in good condition and annual cleaning the roof gutters and checking around the external openings (Efficiency Energy for EU cultural heritage 2013; Sedovic and Gotthelf 2011; Verbeeck and Hens 2005; Hwaish 2015). Two remarkable interventions are shown in the following section:

(1) **Improve the Ventilation of the Indoor Spaces**

The ventilation methods can remove combustion gases and reduce the moisture from the spaces. Natural ventilation is the common and traditional method of providing fresh air and improves the indoor air quality through the air movement from the structure openings themselves. Natural and passive ventilation methods are efficient and more recommended. A study in Kapedes, Cyprus, tested the impact of three different natural ventilation strategies (full day ventilation, day time ventilation, and night ventilation) on their indoor thermal environment in hot summer period. This study emphasized the positive role of natural ventilation for cooling purpose in the traditional residential buildings and indicated that the night cross-ventilation strategy during the summer is the best option. It reduces peak indoor air temperatures and improves the indoor thermal conditions during the following day (Michael et al. 2017). Another study for the for heritage residential building in Catania, Italy, also emphasized the effective thermal role of the night ventilation more than other passive cooling techniques. It also indicated that combining different techniques such as understanding the thermal capacity of the building envelope, adding insulation layer, using light color material for external walls considering appropriate ventilation strategy are useful to improve the thermal comfort (Evola et al. 2015). In case of the existence of a specific source of pollutants and moisture such as the household cooking and bathing sources or the cases of ineffective natural ventilation, the spot ventilation controls air movement by using localized exhaust fans that can quickly aid to enhance the indoor air quality. When both spot ventilation and natural ventilation together do not meet the building ventilation needs, the whole-structure ventilation strategy could be one of the expensive solutions. Whole-structure ventilation systems use fans and duct systems to exhaust stale air and/or supply fresh air into the structure. They provide controlled, uniform ventilation throughout the structure.

The measures of enhancing the indoor air quality of the existing structures rely on the building usage types and using times such as the residential, educational, commercial and official use. The indoor air quality in the night hours of the bedrooms and in day hours for the living room is required, where CO_2 levels are generally lower in summer than in winter. On the other hand, the daytime ventilating in the kitchens is important to remove the polluted gases and moisture. The ventilation rates and CO_2 concentrations become more important in the educational and public buildings where a relatively few m^2 per person in the room such as in the schools, kindergartens, and offices. The window opening area in relation to the number of students and the floor area are the key design parameters. In the office buildings, the internal loads of the occupants' density and office equipment are higher than in other buildings. Additionally, the occupants spend much time in the specific location and their dress code is more formal than in other buildings. More complex requirements to control air quality cooling and providing fresh air could be designed with natural, mechanical, or hybrid ventilation.

(2) **Windows Upgrade**

Restoration of the existing structures relies on the building usage and can be carried out at various levels starting from the renovation of one room and ending with the whole-structure envelope taking into consideration enhancing energy performance as part of the restoration. The renovation includes the windows' replacement, improving the airtightness of building envelope, insulation, as well as an interior upgrade to improve the indoor environment, improving the air flow and the indoor air quality. Maintaining and repairing the historic windows is an important key preservation strategy (Efficiency Energy for EU Cultural Heritage 2013). Reducing solar heat gain, adding human comfort and convenience, and retrofitting the historic windows to improve energy performance are also required to adapt the era challenges. If the windows are old and not original, their material quality determined the importance of its renovation or maintenance. The proper maintenance would almost certainly outlast windows built of new wood or other modern materials. The arched or curved windows of unique glazing panes or decorative patterns or other distinctive features should be preserved. The decision of window repair or replacement must save the historic character and the neighbor historic fabric, where its replacement will impact the historic structure itself and the experience of other people in your community (Preservation Pennsylvania Serves 2013). To avoid the consequences of establishing sustainability policies and window replacement which Europe currently suffers from, the building advantages and characters to do the enhancement and the developing should be considered, where the passive strategies are the best solutions and key conservation strategies. Windows can be repaired, but sometimes that still does not accomplish the property owner targets. According to Preservation Pennsylvania's research, the most common reasons for replacing historic windows with new ones include aesthetics, safety, comfort and convenience, and energy efficiency (Preservation Pennsylvania Serves 2013). Most basements and unopened rooms in the historic buildings have a damp and moldy smell which causes polluted indoor air. Since the historic gallery and basements provide a cost-effective space for reusing and operating that makes improving the ventilation, using the exhaust systems is a significant priority to reduce the potentially harmful effects of gases and allergens and make the space healthier, safer, and eco-friendly.

1.3.4 People Action and the Participatory Among the Action Levels

① **People Action Level**

The significant role of climate in the history of people was a great determinant in many battles and has its effects on the architecture and human activities such as food and clothing. No matter, the narrative of the flood and the Seven Years' Famine in Ancient Egypt is kind of mythology or reality, as well as the frost and the extreme weather in France during the Louis XIV period which caused the famine and many victims due to the cold spring and the rot of wheat. Several historical stories and pieces of evidence in past times emphasized that the people always do the urgent proper adapted action to face the climate problem and save themselves in the catastrophic events. In past times and before the Industrial Revolution and the invention of HVAC systems, the people adapted to the climate by their traditional culture and controlled their indoor microclimate by the hearths and the traditional architecture techniques. The indoor microclimate of buildings was much more dependent on external climate. Modern heating, ventilation, and air control (HVAC) systems had a negative role in changing people climatic adapting and behavior and contributed in creating the global architecture far from the local environment needs and effects.

Nelson et al. defined the resilience as the ability of the technological and governmental policies to handle the society and loading the capacity to adapt to extreme undesirable changing circumstances (Nelson et al. 2007). Other researchers found the people power and capacity of facing the problems are the bases of improving the adaptation process (Berkhout et al. 2006; Walker et al. 2002). Water resilience is defined as all the climate change conditions which are related to the water bodies and system, such as the extreme storms, rainfall, rising water level, flooding, where resilience is connected basically to the social science (Sanders and Timmeren 2018). Some cities had a learned lesson in fast recovering from hurricane and flood disasters due to their people action and behavior during and after the problem more than technology or government measures, such as the American New York city after Sandy hurricane in 2012. On the contrary, New Orleans after Katrina hurricane of 2005 still needs recovery till now (Ovink 2014). The resilience of citizens of a county has extreme storms and floods, or of a country most of its areas under the sea level like the Netherlands, forms a guided lesson and a significant base of the sustainable city development and societal responsibility. The Dutch struggled experience in defense system and water barriers inspired several threatened places over the world. The volunteer participation behavior in emergency circumstances even if it does not historically exist in the urban area could be educated and created, where people cooperation and action become faster, cheaper, and more efficient than government action and measures during the crisis.

Due to the long Dutch history of flooding and dikes problems, the Dutch people used to make the proper behavior and to take their responsibility. The combinations of governmental water authorities and civilian's action taking helped to overcome these disasters. The notable example of people power and cooperation is in the security region of "Zaanstreek-Waterland" North of Amsterdam in the Netherlands, which had a historical existence of the flooding risk that leads to having a self-comprehensive safety managing plan in dynamic cooperation among the civilians, municipalities, and the related authorities and emergency police for help and ambulance. In 2016, the national government used television drama series for raising people awareness about their personal risks with an open Web site for providing data on the predicted risks on the individual houses' Web site. The national government made a comprehensive report about the flood, heavy rainfall, the rising sea level, maintaining lands, and the common people. Several Web site pages also explained the people role and responsibilities, and the role of the stimulating water storage by the gardening inside the cities and the role of the water safety programming (Vergouwe 2016; Sanders and Timmeren 2018).

Dutch Sanders and Timmeren explained the differences among three groups of citizen civil corporation as a passive citizen who accepted the government advice and leadership, passive citizen how to join and support the government work, and active citizen who took the responsibility with the government (Sanders and Timmeren 2018). According to their research, the collective action of official authority and civilian action is rarely successful. The best models of cooperation are when the specialists and civilians are leading the initiative, where the official authorities need more time to start acting after focusing on rescue operations and taking away the threat that makes the division of the action not clear and ineffective for both (Sanders and Timmeren 2018). Changing the actual adjusting behavior of the occupants by using proper control of heating and opening windows has the highest possibility for energy consumption. The locals and occupant behavior and expectation need to be addressed through deeper occupant engagement at all stages; therefore, the occupants have a better understanding of the performance expectations and running off his property (Gupta and Gregg 2015).

② **The Participatory and Cooperation among the Action Levels**

The old city is not a museum; it is a subject to the forces of change. The change is presented by the changes in technology, such as communication and transportation, the changes in economic conditions, the shifts in production, and the distribution of wealth, as well as the changes in the environment such as the influence of pollution and the effects of climate change. The cultural evolution creates shifting values and the switch to accommodate social needs such as population migration, changing demographics, quality of life expectations, and the nature of governance. Heritage urbanism has different influences; it starts with the individuals who present the potential for identity and meaning in each of us, and then with the collective where civilization is a communal act that celebrates our collective potential. It characterizes by the continuity where the history and memory are fundamental aspects of the definition of the place, and the change where the environmental influences, technological innovation, economic shits, and culture transformation play a significant role in the heritage settlement (Hoistad 2014). The most effective solutions could be found by the continuous discussion and communication between the actors during the preservation procedure: the architects, conservators, building physicists, and energy experts. The application of energy efficiency in the built heritage in Germany is relying on regions, occupation, and vocabulary value. Several areas use the solar power, while the listed buildings have special policies for establishing values for energy use in these buildings in order to save their authenticity. There are several German active organizations in different governmental and nongovernmental levels concerning the heritage energy and value issues. These organizations include bodies for the legal frame (such as the State Offices for preservation), energy research services and information, crafts education and services, research funding, administrative works, applied and technical research made by research institutes and universities, and bodies for sustainability certification (Kilian 2011).

The three steps in adapting strategy with the climate change are used in the Netherlands "analyze, ambition, act." The international city climate scan is a new low-cost and low-tech methodology developed and easy to apply anywhere in climate adaptation with the participation of public, private, and academic partnership. It started in Rotterdam and then extended to other cities in Europe and Asia. They developed a set of parameters measurements, such as urban heat, urban water quality, urban air quality, urban floods—infiltration capacity, urban plastic waste pollution, and then drew the maps and provided the collected data to help decision-makers to have a fast assessment of the area and try to evaluate the current environment and put the policies and solutions for the area and in the same time raising the awareness of the participated locals to their neighborhood problem (Boogaard et al. 2018). The participatory and cooperation among all the related actors and the action levels in determining the needs, finding solutions, planning to make decisions and then in funding and implementing can help in achieving the targets, effective results, and actors' satisfactions. Exploring the impacts of our behavior to form our ecological footprint can be divided into two parts: the personal behavior and control, and the collective things and human behavior as a result of being citizens of a nation. Changing behavior and learning how to build eco-citizens could help in reducing the energy without having to build new cities (Vale and Vale 2014). For a more balanced assessment of historic buildings, the integration of locals, stakeholders, and laypersons not only is in the usability decision and predicting and assessing the thermal comfort range, but also shares their opinions, ideas, and experiences in the evaluation of the conservation intervention, energy efficiency measures, and expert strategy for their heritage (van der Schoor et al. 2015; Vieveen and van der Schoor 2016).

References

Adgera, W. N., Arnella, N. W., & Tompkinsa, E. L. (2005). Successful adaptation to climate change across scales. *Global Environmental Change, 15*(2), 77–86.

Agnew, N., & Demas, M. (2015). *Principles for the conservation of heritage sites in China*. Beijing: ICOMOS China and the State Administration of Cultural Heritage.

Appendino, F. (2017). Balancing heritage conservation and sustainable development–The case of Bordeaux. *IOP Conference Series: Materials Science and Engineering, 245*(6).

Arnfield, A. J. (2003). Two decades of urban climate research: A review of turbulence, exchanges of energy and water, and the urban heat Island. *International Journal of Climatology, 23*(1), 1–26.

Balson, K., Summerson, G., & Thorn, A. (2014). *Briefing Paper: Sustainable Refurbishment of Heritage Buildings, How BREEAM helps to deliver*. [Online] Available at: http://www.breeam.com/filelibrary/Brochures/Heritage-Sustainable-Refurbishment-v2.pdf. Accessed February 2, 2016.

Bastian, Z., Spiekman, M., & Troi, A. (2014). *Energy retrofit of cultural heritage buildings*. [Online] Available at: http://www.rehva.eu/fileadmin/REHVA_Journal/REHVA_Journal_2014/RJ_issue_3/P.24/24-27_Bastian_RJ1403_WEB.pdf. Accessed June 1, 2018.

Berenfeld, M. L. (2008). Climate change and cultural heritage: Local evidence, global responses. *The George Wright Forum, 25*(2), 66–82.

Berenfeld, M. L. (2015). Planning for permanent emergency: "Triage" as a strategy for managing cultural resources threatened by climate change. *George Wright Forum, 32*(1), 512.

Berkhout, F., Hertin, J., & Gann, D. M. (2006). Learning to adapt: organisational adaptation to climate change impacts. *Climatic change, 78*, 135–156.

Boogaard, F., Heikoop, R., Bosscher, M., & Akkerman, O. (2018). *Research results of a new method in climate adaptation; The international City Climate Scan*. Groningen, The Netherlands, s.n.

Borges, S., Ehmann, S., & Klanten, R. (2014). *Building better sustainable architecture for family homes*. Berlin: Gestalten.

Brandoni, C., & Polonara, F. (2012). The role of municipal energy planning in the regional energy-planning process. *Energy, 48*(1), 323–338.

Bray, M., Hooke, J., & Carter, D. (1997). Planning for sea-level rise on the South Coast of England: Advising the decision-makers. *Transactions of the Institute of British Geographers, 22*(1), 13–30.

Broersma, S., Fremouw, M., & van den Dobbelsteen, A. (2013). Energy potential mapping: Visualising energy characteristics for the exergetic optimisation of the built environment. *Entropy, 15*(2), 490–506.

Building Futures & ICE. (2010). *Facing up to rising sea levels: Retreat? Defend? Attack?*. s.l.: Building Futures and Institution of Civil Engineers (ICE).

Cassar, M. (2005). *Climate change and the historic environment*. London: University College London Centre for Sustainable Heritage.

Cassar, M. (2009). Sustainable heritage: Challenges and strategies for the twenty-first century. *APT Bulletin, Journal of Preservation Technology, 40*(1), 3–11.

Cassar, M., & Pender, R. (2005). The impact of climate change on cultural heritage: evidence and response. In *ICOM Committee for Conservation: 14th Triennial Meeting The Hague, Preprints*. (Vol. 2, pp. 610–616).

Charter, A. (1931). The Athens Charter for the restoration of historic monuments. In *Athens, 1st International Congress of Architects and Technicians of Historic Monuments*.

Charter, V. (1964). International charter for the conservation and restoration of monuments and sites. In *Second International Congress of Architects and Technicians of Historic Monuments*. Venice: International Council of Monuments and Sites (ICOMOS).

Colette, A. (2007). *Case studies on climate change and world heritage*. Paris: UNESCO World Heritage Centre.

Costanza, R., Kubiszewski, I., Roman, J., & Sutton, P. (2011). *Migration and Global Environmental; DR7a: Changes in ecosystem services and migration in low-lying coastal areas over the next 50 years*. London: UK Government's Foresight.

Creel, L. (2003). *Ripple effects: Population and Coastal Regions*. Washington, D.C.: Population Reference Bureau.

de Almeida, S. L. G. (2014). *Retrofitting and refurbishment processes of heritage buildings: Application to three case studies*. Doctoral dissertation: s.l.

Efficiency Energy for EU Cultural Heritage. (2013). Summary Guide for Local Decision-makers, Technical Guidance on Energy Efficient Renovation of Historic Buildings. [Online] Available at: http://www.3encult.eu/en/deliverables/Documents/WP3_D3.6_20130527_P21_Technical_guide_EE_options_for_local_governments.pdf. Accessed December 16, 2015.

Evola, G., Marletta, L., Costanzo, V., & Caruso, G. (2015). Different stgraffrategies for improving summer thermal comfort in heavyweight traditional buildings. *Energy Procedia, 78*, 3228–3233.

Few, R., Brown, K., & Tompkins, E. (2007). Climate change and coastal management decisions: Insights from Christchurch Bay, UK. *Coastal Management, 35*(2–3), 255–270.

Ford, J., & Ford, L. (2011). *Climate change adaptation in developed nations from theory to practice* (Vol. 42). s.l.: Springer Science & Business Media.

Gálvez, F. P., et al. (2013). Sustainable restoration of traditional building systems in the historical centre of Sevilla (Spain). *Energy and Buildings, 62*, 648–659.

Gauzin-Müller, D. (2002). *Sustainable architecture and urbanism: Concepts, technologies, examples*. s.l.: Springer Science & Business Media.

Gillot, L. (2010). Towards a socio-political history of archaeology in the Middle East: The development of archaeological practice and its impacts on local communities in Syria. *Bulletin of the History of Archaeology, 20*(1).

Gong, X., Akashi, Y., & Daisuke, S. (2012). Optimization of passive design measures for residential buildings in different Chinese areas. *Building and Environment, 58*, 46–57.

Graaf, D. E., et al. (2006). *Floating City Ijmeer: Accelerator for Delta Technology*. s.l.: s.n.

Graeme, E. (2015). Design for hydrocitizens: Architectural responses to the defend-retreat-attack scenario. *Sustainable Mediterranean Construction, 1*(2), 89–92.

Grannis, J. (2011). *Adaptation tool kit: Sea-level rise and coastal land use. How governments can use land-use practices to adapt to sea-level rise*. Washington, D.C.: Georgetown Climate Center.

Guhathakurta, S., & Gober, P. (2010). Residential land use, the urban heat island, and water use in Phoenix: A path analysis. *Journal of Planning Education and Research, 30*(1), 40–51.

Guo, X. (2015). *The evolution of the modern conservation in China*. Chongqing, China: Faculty of Architecture and Urban Planning, Chongqing University.

Gupta, R., & Gregg, M. (2015). Do deep low carbon retrofits actually work? *Energy Procedia, 78,* 919–924.

Hansen, J., et al. (2006). Global temperature change. *Proceedings of the National Academy of Sciences, 103*(39), 14288–14293.

Hendriks, C. F., Bijen, J. M. J. M., & Felix, F. (1999). *Duurzame bouwmaterialen*. Best, The Netherlands: s.n.

HM Government. (2011). *The carbon plan: Delivering our low carbon future*. s.l.: s.n.

Hoistad, M. A. (2014). *Sustainable urbanism: A matter of survival, purpose, ecology and heritage*. Chongqing, China: Faculty of Architecture and Urban Planning, Chongqing University.

Horowitz, A. D. (2013). The effects of sea level rise on historic districts and the need for adaptation. s.l.: MAHP diss. Goucher College.

Horowitz, A. (2016). Planning before disaster strikes: An introduction to adaptation strategies. *APT Bulletin, Special Issue on Climate Change and Preservation Technology, 47*(1), 40–48.

Hwaish, A. N. A. (2015). Impact of heat exchange on building envelope in the hot climates. *International Journal of Emerging Technology and Advanced Engineering, 5*(2).

ICOMOS. (1994). *The Nara document on authenticity*. Nara : s.n.

ICOMOS. (1999). The Burra Charter: The Australia ICOMOS charter for places of cultural significance: With associated guidelines and code on the ethics of co-existence. In: Burwood (Ed.). s.l.: Australia ICOMOS, 2000.

IUCN, UNEP & WWF. (1980). *International Union for Conservation of Nature, Natural Resources, and World Wildlife Fund. World conservation strategy: Living resource conservation for sustainable development*. Gland, Switzerland: IUCN-UNEP-WWF.

Kaiser, R. (2009). Stehen Denkmalschutzauflagen im Widerspruch zur Energieeffiziens? *Die Denkmalpflege, 67,* 69–72.

Kershner, J. (2010). Climate Change Adaptations for Land Use Planners. [Online] Available at: http://www.cakex.org. Accessed March 20, 2017.

Khalili, M., & Amindeldar, S. (2014). Traditional solutions in low energy buildings of hot-arid regions of Iran. *Sustainable Cities and Society, 13,* 171–181.

Kilian, R. (2011). *Energy Efficiency and Historic Buildings in Germany*. [Online] Available at: http://eprints.sparaochbevara.se/663/1/Energy_efficiency_and_historic_buildings_in_Germany.pdf. Accessed June 06, 2018.

Kim, S. A., Ryan, A., & Meashama, T. G. (2012). Managed retreat of coastal communities: Understanding responses to projected sea level rise. *Journal of Environmental Planning and Management, 55*(4), 409–433.

Klein, R., Smit, M., Goosen, H., & Hulsbergen, C. (1998). Resilience and vulnerability: Coastal dynamics or Dutch dikes? *Geographical Journal, 164,* 259–268.

Kourgialas, N. N., & Karatzas, G. P. (2014). A hydro-sedimentary modeling system for flash flood propagation and hazard estimation under different agricultural practices. *Natural Hazards and Earth System Sciences, 14*(3), 625–634.

Lee, Y. (2014). Coastal planning strategies for adaptation to sea level rise: A case study of Mokpo, Korea. *Journal of Building Construction and Planning Research, 2,* 74–81.

Linnean Solutions, The Built Environment Coalition, The Resilient Design Institute. (2013). Building Resilience in Boston. "Best Practices" for Climate Change Adaptation and Resilience for Existing Buildings. [Online] Available at: https://www.cityofboston.gov/images_documents/Building_Resilience_in_Boston_FINAL_tcm3-40185.pdf. Accessed July 7, 2015.

Locke, W. J. (1904). *Recommendations of the Madrid Conference (1904). Vol. XI. Third Series, Point 2* . The Architectural Journal: Journal of the Royal Institute of British Architects (RIBA), Volume XI. Third Series, point 2.

Magrini, A., & Franco, G. (2016). The energy performance improvement of historic buildings and their environmental sustainability assessment. *Journal of Cultural Heritage, 21,* 834–841.

McCarthy, J. J., et al. (2001). *IPCC Climate change 2001: Working group II: Impacts, adaptation and vulnerability. Summary for Policymakers, 2.7'*, Cambridge, UK: Intergovernmental Panel on Climate Change, Cambridge, UK: Intergovernmental Panel on Climate Change, http://www.grida.no/climate/ipcc_tar/wg2/index.htm.

Michael, A., Demosthenous, D., & Philokyprou, M. (2017). Natural ventilation for cooling in mediterranean climate: A case study in vernacular architecture of Cyprus. *Energy and Buildings, 144,* 333–345.

Mohamed, N. A. G., & Ali, W. H. (2014). Traditional residential architecture in Cairo from a green architecture perspective. *Arts and Design Studies, 16,* 6–26.

Moschella, A., et al. (2013). Historic buildings in mediterranean area and solar thermal technologies: Architectural integration versus preservation criteria. *Energy Procedia, 42,* 416–425.

Nakicenovic, N., et al. (2000). *Special report on emissions scenarios: A special report of Working Group III of the Intergovernmental Panel on Climate Change*. Cambridge; New York: Cambridge University Press.

Middel, A., Häb, K., Brazel, A. J., Martin, C. A., & Guhathakurta, S. (2014). Impact of urban form and design on mid-afternoon microclimate in Phoenix Local Climate Zones. *Landscape and Urban Planning, 122,* 16–28.

Nelson, D. R., Adger, W. N., & Brown, K. (2007). Adaptation to environmental change: Contributions of a resilience framework. *Annual Review of Environment and Resources, 32.*

New, M., Liverman, D., Schroeder, H., & Anderson, K. (2011). Four degrees and beyond: The potential for a global temperature increase of four degrees and its implications. *Philosophical Transactions of the Royal Society A, 369,* 6–19.

Nicholls, R. (2011). Planning for the impacts of sea level rise. *Oceanography, 24*(2), 144–157.

Ovink, H. (2014). *Rebuild by design: Recovering New York after Sandy*. s.l.: s.n.

Palmer, J., & Cooper, I. (2013). *United Kingdom housing energy fact file 2013.* In s.l.: Department of Energy & Climate Change, Prepared under contract to DECC by Cambridge Architectural Research, Eclipse Research Consultants and Cambridge Energy. The views expressed are not necessarily DECC'sp 1.

Pearson, M., & Williams, E. (1996). Possible effects of climate change on the cultural National Estate. *Tempus-St Lucia: Anthropology Museum, University of Queensland, 5,* 122–126.

Peel, M. C., Finlayson, B. L., & McMahon, T. A. (2007). Updated world map of the Koppen-Geiger climate classification. *Hydrology and Earth System Sciences Discussions, 4*(2), 439–473.

Preservation Pennsylvania Serves. (2013). *Preservation Pennsylvania, Considering the Repair, Retrofit, and Replacement of Historic Windows.* [Online] Available at: http://preservationpa.org/uploads/2016-Preservation-Pennsylvania-Windows.pdf. Accessed May 04, 2018.

Prizzia, R. (2015). *Climate change and disaster management.* s.l.: Sentia Publishing.

Protocol, K. (1997). *Article 3 of the UN framework convention on climate change.* Kyoto, Kyoto Protocol.

PVSITES. (2016). *European climate zones and bio-climatic design requirements*, s.l.: PVSITES project partner BEAR-iD, with additional contributions from NOBATEK.

Rabah, K., & Tamkan, Z. (2002). Application of ARCHIPAK for development of passive solar energy-efficient building in North Cyprus. *Architectural Science Review, 45*(3), 219–229.

Rahman, S., & Rahman, M. A. (2015). Climate extremes and challenges to infrastructure development in coastal cities in Bangladesh. *Weather and Climate Extremes, 7,* 84–95.

Rahmoun, T., Hassan, M., & Alhasan, W. (2016). Protection strategy for the coastal areas of climate change "Syrian Coast, Tartous City". *Research Journal of Applied Sciences, Engineering and Technology, 12*(3), 264–271, ISSN: 2040-7459.

Rappoport, P. (2015). Best practice in cultural built heritage. [Online] Available at: http://www.heritage21.com.au/best-practice-in-cultural-built-heritage/#. Accessed November 18, 2017.

Rijcken, T. (2003). Neerlands H2oop, voorzieningen voor waterwijken. Delft, the Netherlands: s.n.

Roggema, R. (2009). *Adaptation to climate change: A spatial challenge* (Vol. 6). s.l.: Springer Publisher.

Ruskin, J. (1885). *The seven lamps of architecture* (Vol. 7). New York: John B. Alden.

Sabbioni, C., & Bonazza, A. (2010). How mapping climate change for cultural heritage? The Noah's Ark Project. In R.-A. Lefevre & C. Sabbioni (Eds.), *Climate change and cultural heritage* (pp. 37–42). Edipuglia: Bari.

Sabbioni, C., Brimblecombe, P., & Cassar, M. (2010). *The atlas of climate change impacts on European cultural heritage: Scientific analysis and management strategies*. London: Anthem Press.
Saleh, H. S., & Saied, S. Z. (2017). Green architecture as a concept of historic Cairo. *Procedia Environmental Sciences, 37,* 342–355.
Sanders, F. C., & Timmeren, A. V. (2016). *'Zero Transition' in housing areas prosper by balancing municipality-control and citizens-participation, by bundling of Dutch IKS2 case-study results*. Utrecht, The Netherlands: SBE16 Sustainable Built Environment 2016—Transition Zero.
Sanders, F. C., & Timmeren, A. V. (2018). *Dutch circular cities by the energy of people: Post PhD best practices research on Amsterdam and Rotterdam citizen initiatives*. Abu Dhabi: Construction on Urban Future.
Sedovic, W., & Gotthelf, J. H. (2011). *Paradox to paradigm: Sustainability & performance of heritage buildings*. Dubrovnik, Croatia, s.n.
Shalkany, A. H. (2014). *Environmental values of heritage architecture in modern Egypt during the 19th and beginning of 20th centuries—Cairo city as a case study*. s.l.: Masters thesis of Science in Architecture Cairo University, Faculty of Engineering, Cairo, Egypt.
Sharifi, A., & Murayama, A. (2013). A critical review of seven selected neighborhood sustainability assessment tools. *Environmental Impact Assessment Review, 38,* 73–87.
Sköld, Y. A., et al. (2015). An integrated method for assessing climate-related risks and adaptation alternatives in urban areas. *Climate Risk Management, 7,* 31–50.
Smit, B., Burton, I., Klein, R., & Wandel, J. (2000). An anatomy of adaptation to climate change and variability. *Climatic Change, 45,* 223–251.
Solomon, S., et al. (2007). IPCC Fourth Assessment Report: Climate Change. *Contribution of working group I to the fourth assessment report of the intergovernmental panel on climate change*. Cambridge, UK and New York, NY: Cambridge University Press.
Tomás, N., Carvalho, A., & Coelho, D. (2010). *Renewable Energy Integration in Buildings: A Case Study in Portugal*. Granada, Spain, s.n.
Troi, A. (2011). *Historic buildings and city centres—The potential impact of conservation compatible energy refurbishment on climate protection and living conditions*. Dubrovnic, s.n.
United Nations. (1987). Our Common Future—Brundtland Report., s.l.: United Nations.
Vale, B., & Vale, R. (2014). *Eco-cities or Eco-citizens? A discussion of cities, buildings and users*. Chongqing, China: Faculty of Urban Construction and Environmental Engineering, Chongqing University.
van den Dobbelsteen, A., Keeffe, G., Tillie, N., & Roggema, R. (2010). Cities as organisms: Using biomimetic principles to become energetically self-supporting and climate proof. In *Proceedings of the First International Conference on Sustainable Urbanization, Hong Kong, China.*
van der Schoor, T., Acevedo, E. C., & Vieveen, M. (2015). Valuation of medieval churches; towards an integration of experts' and laypersons' views.
Verbeeck, G., & Hens, H. (2005). Energy savings in retrofitted dwellings: Economically viable? *Energy and Buildings, 37*(7), 747–754.
Vergouwe, R. (2016). *The water safety of the Netherlands mapped (Dutch: Deveiligheid van Nederland in kaart)*. The Netherlands: Rijkswaterstaat Ministery M&I.
Vieveen, M., & van der Schoor, T. (2016). Integral design method for Energy efficient Restoration. In *Proceedings EECHB 2016* (pp. 318–324).
Wai-Yin, C., & Shu-Yun, M. (2004). Heritage preservation and sustainability of China's development. *Sustainable Development, 12*(1), 15–31.
Walker, B., et al. (2002). Resilience management in social-ecological systems: A working hypothesis for a participatory approach. *Conservation ecology*, s.l.: s.n.
Wamsler, C., Brink, E., & Rivera, C. (2013). Planning for climate change in urban areas: From theory to practice. *Journal of Cleaner Production, 50,* 68–81.
Winter, T. (2014). Beyond Eurocentrism? Heritage conservation and the politics of difference. *International Journal of Heritage Studies, 20*(2), 123–137.

Chapter 2
Evaluation and Management Approaches

In response to the urgent need to address the new challenges facing conservation practices in the world, several integrated approaches and preservation studies worldwide in partnership with the community and other stakeholders could help in managing the cultural heritage. This chapter highlights some important steps, methods, and theories could be used in monitoring and management the built heritage. It started with the advanced leadership and enhancement visions of macrolevel management approaches that work hand by hand with the documenting, value, diagnosis, and problems identification, then the importance of using simulation programs as assessment tools. The common evaluation and assessment approaches and international standards for individual heritage structures and usages in term of thermal comfort concepts have been also summarized and presented. Special considerations need to be taken in suggesting the appropriate usage of individual historic buildings. Improvements considering the building envelope and value are helpful to adapt to the indoor environment for the occupants and to increase productivity during the operating time. Field measurements and survey, thermal comfort and PMV–PPD adaptive models, socioeconomic and cultural issues should also be considered.

2.1 Monitoring and Management Frameworks of Heritage Sites and Buildings

Since late nineteenth century, the definition of heritage gradually changed from monuments to historic cities in the 1960s, to cultural landscapes recently, thus emphasizing the relationship between territory and man presence. Urban heritage protection dealing initially with the physical conservation of historic cities was extended in the 1980s into functional conservation, with the aim to counteract gentrification, and since 1990s, it is principally focusing on the protection of "place identity" of historic cities, often endangered by mass tourism and the transformation of heritage sites. In 2002, the Budapest Declaration switched the emphasis from the protection of the

M. Hassan and H. Xie, *Climate Change and Conservation of Coastal Built Heritage*, https://doi.org/10.1007/978-981-13-8672-5_2

historic sites to management, and UNESCO invited listed WHS to prepare management plans. Since 2006, emphasis is placed on monitoring and on the impacts created by the UNESCO World Heritage status, passing from management to monitoring (Protection-to-Management–Monitoring) (Porfyriou and Re 2015). It is necessary to start with the development of comprehensive plans for the management of archaeological and natural sites according to the monitoring methodology, including phase plans for management according to sequential stages. Each phase is evaluated using specific indicators whose values range from a starting point to the optimal situation to end with a critical situation to give the final judgment and decision-making. Then examine the impact of management of these sites on the property itself and on both surrounding land and the community, to become mandatory management plans in subsequent stages.

The modes of strategic thinking and the advanced leadership such as Zoom Out and Zoom In and Top-Down and Bottom-Up Processing linked to the time provide a hierarchical structure of safe and resilient conservation actions (Kanter 2011). Top-down and bottom-up are two approaches for planning and ordering the management works processes. Top-down approach starts with a comprehensive view of the whole projects with the basic plans, guide policies, and authority decisions and then goes down into smaller aspects and details and implementation roles. Top-down measures have long been the core of conservation policies when there is a need for strong implementation and protection by using regulations to force the actors to do at least the basic advanced objectives. The bottom-up approach starts with data and details of the small levels of the project, then organizes and gathers all the aspects to form more complicated level or the final product and to implement the project objectives and basic decisions. The bottom-up approach is the logical alternative to the top-down because it addresses the details and many small factors that create the whole image of any project. Mixed approaches of up and down action would ensure the virtues of both perspectives and to create the ability to move across a continuum of perspectives in both directions to face the problems of climate change, and down to find the solutions for local people needs and provide a whole image perception, Fig. 2.1.

The World Commission for Protected Areas (WCPA) developed a framework for monitoring and management the heritage sites and urban areas. WCPA framework provides a cyclical process of assessment, consisting of six phases divided into twelve

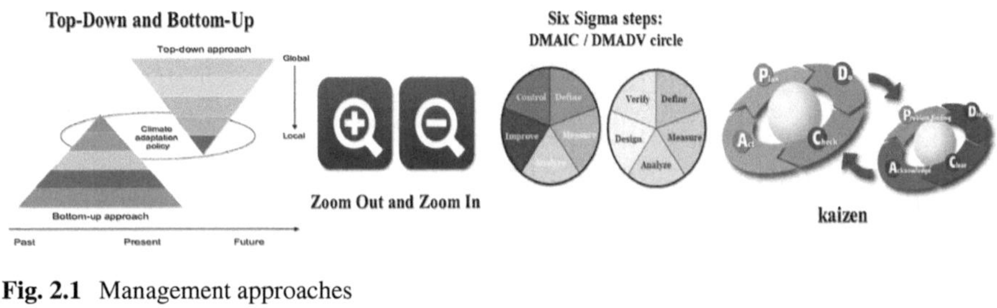

Fig. 2.1 Management approaches

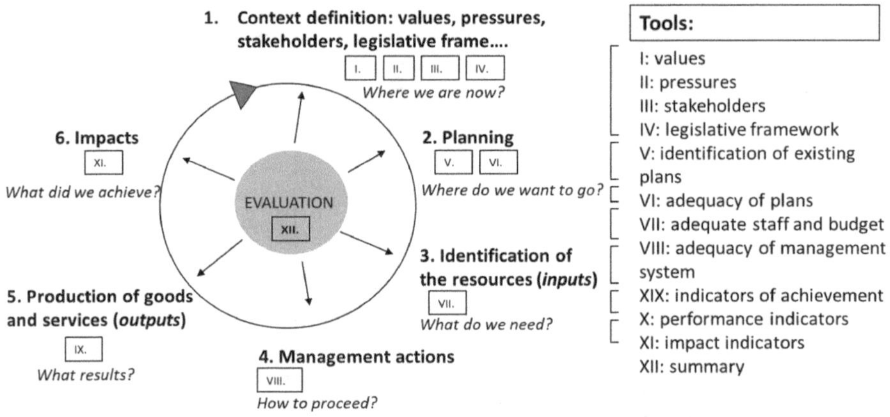

Fig. 2.2 WCPA framework. *Source* Porfyriou and Re (2015)

tools. The phases include the context definition (values, pressures, stakeholders, legislative frame, etc.), planning, identification of the resources (inputs), management actions, production of goods and services (outputs), and impacts, as shown in Fig. 2.2.

On the other hand, Six Sigma and Kaizen approaches could be used as a process management approaches to diagnose and improve the built heritage management. They guide the project through the steps of improvement, from problem pinpointing to the implementation of result/solutions into the management system. The Six Sigma methodology is defined by two five-step circles: DMAIC circle (Define, Measure, Analyze, Improve, and Control) or DMADV circle (Define, Measure, Analyze, Design, and Verify). In addition to the five steps, there is also a step zero that occurs first. It is known as Six Sigma leadership. This approach based on data collection and measurements to predict the results and for improve/design and control/verify the projects based on solid analysis and data. This approach needs professional teamwork of experts, leader, and stakeholders. The Six Sigma approach is a structured, disciplined, rigorous approach to process improvement, where each phase is linked logically to the previous as well as to the next phase in the aim to reduce the volume of mistakes during the conservation process (Rath and Strong 2002; Dumitru and Dumitru 2008). The Japanese approach for the continuous improvement "改善:Kaizen" is suitable for well-defined, relatively direct and simple problems that do not involve numerous functions or complex processes, or when the dissatisfactory performance of the current state is due to only a few factors that do not vary widely over time. The continuous cycles of Kaizen activity include PDCA (Plan → Do → Check → Act). The format for Kaizen can be individual, suggestion system, small group, or large group. It is a daily process often used to focus on continuous small improvements where the operators mostly look for small ideas for the details of the implementation as a way of fast action to feel the gap between the theory and practice and solve the problem to increase the productivity. The successful implementation requires the participation of people at all levels of an organization in the improvement. The aligned small improvements and standardization of the persons, small group, or large group yield large contribution in terms of overall improvement in productivity under the small line key objective.

2.1.1 Documenting, Diagnosis, and Simulation Programs

Conservation of heritage sites in the Principles China 2002 involves six steps undertaken in the following order: identification and investigation; assessment; formal proclamation as an officially protected site and determination of its classification; preparation of a conservation master plan; implementation of the conservation master plan; and periodic review of the master plan (Barker et al. 2002). The conservation process starts with the identification and documenting then going down to other steps. The documenting of the cultural heritage includes a mechanism for recording the material so to build databases as to preserve it and make it available for the scientific researchers to find the proper ways of protection and developing of our heritage, and to ensure the continuity of our cultural heritage asset. Furthermore, the mechanism is significant to make the databases available for others and passing within the local communities for the locals themselves and future generations. It has a significant role as a resource in the safeguarding of the tangible and intangible cultural heritage, helping in establishing the legalization and the preservation actions and options. The documenting includes the written descriptions, drawings, digitizing the ancient manuscript, writing down spoken traditional stories, photographing, audio recording, filming, survey, measured data, etc. where the new technologies provide many opportunities of documenting and digitizing the cultural heritage and pass on to succeeding generations (UNESCO 2003). The tangible and intangible cultural heritage and the traditional knowledge and expressions reflect the communities' identities. Therefore, the safeguarding of them is a duty for many stakeholders. They take part in documentation and have different needs and roles in the managing and preserving. The museums, archives, libraries, universities, municipalities, and other cultural institutions, which specifically deal with the management of cultural heritage assets, participate in recording the cultural heritage hand in hand with the local community where the benefits of documentation are in the interests of both holders and the public such as the legal or policy advisors, research, educational or cultural institutions, private sector partners, and government and public sector agencies.

Diagnosis aids in finding building defects such as damp, insect infestation, mold growth and the structural building failure of building materials and envelope elements and external finishes. Building pathology is a building's analyzing methodology and investigations which determines the defects, durability and performance of the buildings and their environments to help create the right remedial and risk management resolutions, for preventative maintenance plans, future repairs, alterations and renovations to be done in a lower cost and academic manner reserving the unique features of the historic and listed buildings (Van den Beukel and CIB 1993). As in the medical examination where the patient is clinically examined, the historic structure is visually examined, as well as by using the measuring equipment and often according to a checklist or following a specific methodology to confirm the type of problem in order to later treat the causes, or finding ambulance solutions and increase the age opportunities for a longer time. Unfortunately, the examination in most cases is not done for preventive purposes, but for emergency solutions, or for renovations, reuse or refurbishment purposes. Diagnosis report provides a detailed knowledge of the buildings construction, usage, and maintenance and the various mechanisms which affect their structural, material, and environmental conditions. It typically begins with a compilation of background and historical information, nondestructive investigation techniques, as well as specialist instrumentation and laboratory facilities (Watt 2009).

Simulation asses and tests the solutions without any physical risks and facilitates the right decision-making and safe implementation. Currently, the modeling and configuration programs help in investigating the current situation of the case study site and the proposal scenarios to find effective strategies. Numerous computer programs (such as ENVI-met modeling, CFD, Phoenix, and ECOTECT) have been developed for design, assessment, proactive building simulations software, simulations supported by statistical analysis to identify correlations and interdependencies between inputs, and evaluating the performance which aid decision-making in early design and provide several benefits of energy analyses, equipment sizing, and predicted indoor comfort requirements (Østergård et al. 2016). The process of the preservation consists of the preservation of cultural heritage in a fixed form, such as when they are built or documented. Then try to have added value using the framework for developing, passing on, and governing access. Monitoring the structural health of the historic buildings is a daunting task due to the lack of previous studies and information. The main goal of the monitoring is helping the maintenance and restoration works and predicts the structures performance and reaction to the unpredicted stresses. The vocabularies' documenting, recording, and digitizing are developing practical tools for managing the cultural heritage action and options and form a guide to create a resilient historic building project. In particular, the heritage in danger or needs to be further documented for a specific purpose. The community-led documentation, strengthening the capacity, and promoting the interests of the cultural heritage holders are significant issues to preserve their heritage for themselves and for future generations.

2.1.2 Significance and Values Evaluation

The heritage buildings have a significant value. This value can be architectural, artistic, historic, social, or environmental or full complement value. It differs from site to site and from block to another one. All the properties inscribed on the world heritage list have the common feature and significance which evaluated by the UNESCO criteria for outstanding universal value (OUV) (Rappoport 2015). Regardless of where the world heritage properties are located, they are important to and belong to everyone, and they are forming an irreplaceable legacy that the global community has decided to protect for the future and for the next generations. The listed built heritage and the monumental structures have the priority to preserve and protect more than other sites. Several historic buildings have a historical value, and significant features are not formally listed and protected; therefore, many approaches to evaluating the built heritage in the context of heritage place protection have been used. The monitoring of the heritage buildings and sites is an essential part of preserving their values and developing possible techniques to save their structures and environments (Mason and de la Torre 2002). The principles for the Conservation of Heritage Sites in China (2015) determined the value types and the significance of its assessment as a first step of the conservation process. The evaluation of the heritage values approaches based on several historic, artistic, scientific, cultural, economic, and social and educational values (Mason 2002; Allen Consulting Group 2005; Agnew and Demas 2015). The assessment could be started with the grounded significance of witnessing history, artistic, and aesthetic preference, scientific or technical creativity evidence, and feats. As well as, with the value cultural values which include several points, firstly, the identity value based on the emotional relationship between the society and the site and on the influences of a resource. Secondly, the artistic or technical value based on the structure significant design itself, research, and influences of the resource treatment. Thirdly, the authenticity and the rarity of the resources based on calculation and the level of the resource protection, and the physical and visual value and the impact on the viewer. The economic values of the heritage based on the concept of the heritage as a cultural resource for economic benefits, marketing, and outcomes and influence community priorities and future conservation funding from tourism, commerce, use, and amenities where the economic value of the cultural heritage can be classified into two main groups based on the usage: use value (direct and indirect use value and their benefits) and non-use value and the willing of the people who make direct or indirect use of the place to share in protecting their heritage (Frey 1997). The social values and the potential for social interaction and establishing community identity based on the assessment and influences interpretation and presentation of the site and visitor interaction. As well as the educational values and the extent to which a resource can inform the present about its past in the context of cultural tourism. It is based on the assessment of physical integrity and historical research and influences reuse and interpretative decisions. We refer two cases, for example, but are not limited to. The evaluating of the German historic monument includes six levels of values that have to be taken into account through the preservation measures. These levels include the urbanistic value for a townscape or landscape, the value for settlement or social history, the artistic quality of all components of the architecture, the quality of the interior, the value as a source for heritage science, and marks and evidence of history. The energy efficiency measures related to the last three values of heritage vocabulary such as walls insulation, plaster, openings, and lighting (Kiesow 1982). Additionally, according to the National Board of Antiquities (NBA), which is the authority responsible for defining the cultural and historical value of buildings in Finland, the buildings of particular historical and cultural value in Finland were listed as follows (Nypan 2009; Ronchi and Nypan 2006):

(1) Historic buildings protected by town plan,
(2) Conservation areas protected by town plan,
(3) Buildings, monuments and sites protected by Building Protection Act, Church Act, Antiquities Act, Decree on the Protection of State-owned Buildings,
(4) Buildings, monuments, and sites for which historical and cultural value has been recognized in national, regional, or local inventories,
(5) Monuments and sites submitted to the list of UNESCO Nature and Cultural Heritage List,
(6) Buildings subsidized by the National Board of Antiquities or Regional Environment Centers due to their cultural and historical value,
(7) On a case-by-case basis, other buildings and monuments being of particular cultural and historical value considered equal to the buildings included in the points 1–6.

2.2 Thermal Comfort and Building Usages

The comfort of the occupants of a building is the combination of thermal comfort, visual comfort, acoustic comfort, etc. where the thermal comfort is considered as the most significant (UNI EN ISO 7726 2002). Thermal comfort in the indoor environment of the historic buildings is an important issue for the occupants' health and preserving the place identity and the continuity of the local architectural heritage and the cultural role of the old cities and maintains their human resources and historic value. Overheating problems of the hot period may affect the occupants' satisfaction and have a direct influence on the energy usage for the air-conditioning uses, so it raises the use of fossil fuel, then lead to a rise in the atmospheric pollution and finally climate change (Hassan et al. 2018). The introduction of heating, ventilation, and air-conditioning (HVAC) systems moved people away from passive design and rapidly increased energy demand and consumption. Recent studies have encouraged natural and passive systems and tend to combine passive and active design systems to achieve maximum energy efficiency in the building sector. Studies have also considered enhancing the thermal comfort and energy performance of all building sectors, including a high percentage of existing and historic buildings (Antonio et al. 2016). The interventions in historic buildings face many challenges and constraints. The usage requirements, operation, maintenance, and cost should be carefully defined. In many cases, the historic building is reused based on the investment objectives of the decision-makers without much consideration of the building's character and the end users' needs. Evaluating the passive thermal behavior of the building envelope and the thermal properties of the various components, such as the construction material and wall system, is necessary (Atsonios et al. 2013).

Thermal studies yielded two thermal comfort models—heat balance and adaptive. The early model used the principle of human thermal balance with his environment condition to keep his body in a constant internal temperature. This model used the climate chamber experiments to examine the people responses (skin temperature and sweating) during doing their activities. Fanger developed a comfort equation to predict neutral conditions (Fanger 1972; Van Hoof 2008). The heat balance model is intended for the constant and static conditions of the air-conditioned indoor environment and ignored the occupants' adaptation behavior. The adaptive models have mainly based on field measurements and surveys data in real living and working environments. People broadly interact with their spaces' environment and tend to restore their thermal comfort by changing their behavior and through their physiological and psychological adaptation (Roaf et al. 2010). People social and cultural traditions, behavior, clothing, activities level, opening/closing windows, and their proper time to switch on the fans and mechanical systems are influenced by their perceptions of the surrounding environment. The adaptive model studies also found the influence of the outdoor temperature and the indoor neutral temperature. People interaction and creation the indoor environmental systems form the mechanism of achieving thermal comfort as shown in Fig. 2.3. After the latest observed serious heat waves and the expected rise in temperature in the future due to climate change impact, the international and local calls emerged to minimize the financial and ecological expenses of energy consumption and encourage the energy-saving measures and passive cooling strategies by using the natural ventilation to raise the indoor airflow and enhance the thermal comfort without having any adverse effects on the occupants' comfort and life quality.

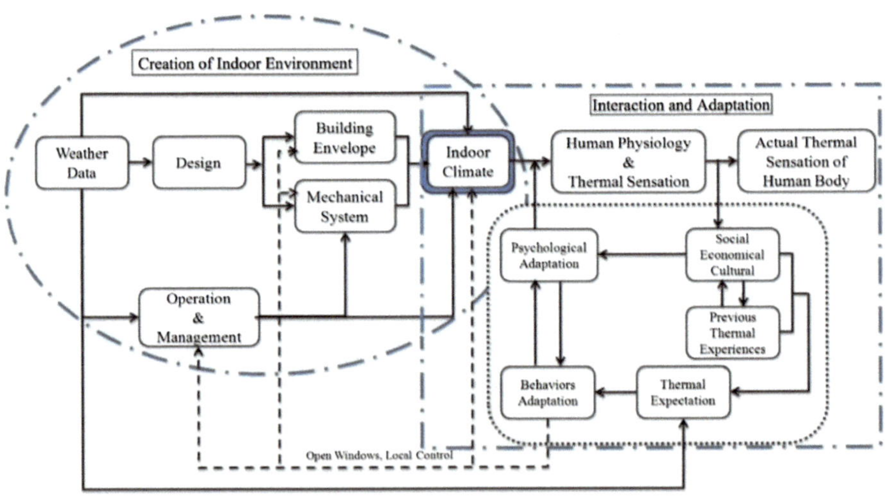

Fig. 2.3 Adaptive thermal comfort mechanism. *Source* Li et al. (2012)

2.2.1 *Thermal Requirements for the Common Usages*

The HVAC systems completely separate the function of an envelope from the delivery of a comfortable environment. Therefore, the recent research call for the environmental strategy is compatible with climate change and rising temperature away from the buildings that use air conditioners to create a comfortable interior air and reducing the excessive use of energy in order to achieve the desired temperature in the interior spaces in summer and winter. The measures of enhancing the indoor air quality and the thermal comfort of existing structures rely on building usage and types such as the residential, educational, and commercial and official use. In our research, the thermal comfort in the common and important functions in the historic buildings in each of houses, offices, and museums would be introduced where the museum environment deals with three main issues visitors comfort and proper environment for the collection, and building materials at the same time. The other functions such as commercial, restaurants and the other entertainment and services functions are not our interest in this research due to:

- The short stay duration for the occupants in this kind of building,
- The need for proper heating and cooling system in this kind of building to provide the occupant the comfort during his short time,
- The small amount of these kinds of building usage compared to the percentage of the residential and official uses.

① Indoor Thermal Requirements for the Official Usage

The high values of the humidity cause discomfort feelings, excessive perspiration, exacerbation of the effects of high temperature, etc., and the low values can cause respiratory problems. Optimum humidity levels are between 40 and 60%, and they could be extended to be between 30 and 70%, where the humidity levels below 40% will begin to cause problems for workers with conditions such as sinusitis. Table 2.1 shows the temperature and humidity ranges for the office comfort in summer and winter as recommended by the thermal environmental conditions for human occupancy which established by the American Society of Heating, Refrigerating, and Air-Conditioning Engineers (ASHRAE) Standard 55-2010 and the Office Ergonomics of the Canadian Centre for Occupational Health and Safety (Bank 1998; ASHRAE, ANSI 2010).

Table 2.1 Temperature/humidity ranges for comfort

Conditions	Relative humidity	Acceptable operating temperature (°C)
Summer (light clothing)	If 30% then If 60% then	24.5–28 23–25.5
Winter (warm clothing)	If 30% then If 60% then	20.5–25.5 20–24

Source Bank (1998), ASHRAE, ANSI (2010)

Several field studies based on the model of the adaptive thermal comfort have been made in many countries worldwide with the aim of achieving a better understanding of thermal comfort to face climate change and the demands for more energy-efficient buildings and more productivity (Rupp et al. 2015; Lan et al. 2009). In order to verify the applicability of the adaptive model of thermal comfort in mixed-mode buildings, studies were carried out in offices in Shenzhen in China and in Melbourne and Sydney in Australia. The different conditioning modes were perceived differently by users. PMV–PPD model is inadequate to describe the thermal comfort in mixed-mode buildings. The adaptive model is more applicable to this type of building during the use of natural ventilation (Nicol and Humphreys 2007; Luo et al. 2015; Drake et al. 2010).

According to the latest revision of the ASHRAE 55 standard, the application of adaptive comfort theory is limited to pure natural ventilated spaces where occupants can freely adapt to indoor and outdoor thermal conditions through adaptive opportunities such as windows and clothing. Based on the satisfaction and thermal acceptability of the users during summer, the researchers in different climate zones, such as Australia, Malaysia, and China, found that the users of the naturally ventilated offices are more tolerant concerning their thermal environment than the users of the offices with hybrid ventilation, despite experiencing higher temperatures (Deuble and De Dear 2014; Frontczak and Wargocki 2011; Daghigh et al. 2009; Yang and Zhang 2008). Post-occupancy evaluation (POE) methods are used in the USA, Australia, and New Zealand to provide feedback to building managers on the performance of their building after completion, as a means of improving service for future clients (Deuble and De Dear 2014). As building managers and designers continually strive to improve occupant satisfaction and productivity by ensuring comfortable and healthy working conditions, POE represents a systematic quality assurance process toward these ends. Using the post-occupancy evaluation database for offices, Kim and de Dear identified that the type of the offices conditioning influences the expectation of users concerning indoor environment quality satisfaction. The proper thermal conditions improved overall satisfaction with the working environment and had positive effects in naturally ventilated buildings. While in the air-conditioned buildings, the thermal conditions were associated with negative evaluations regarding the overall environment, and thermal conditions provided both positive and negative impacts in mixed-model structures (Kim and de Dear 2012). A study in Cyprus emphasized the positive role of natural ventilation strategies for cooling purpose in the historic buildings in the eastern Mediterranean region in the summer period (Michael et al. 2017). Other research in India highlighted the increased air velocity by fans as one of the measures used by participants to improve the comfort conditions (Indraganti et al. 2014).

② Indoor Thermal Requirements for the Residential Usage

Adaptive behavior and environmental control play an essential role in adjusting to the indoor environment in the houses. Field study in Finland showed that the Finnish respondents had less thermal control and adaptive opportunities in the offices than in houses, and the residents showed better thermal comfort levels (Karjalainen

2009). The residents use different mechanisms of adaptation in their homes such as changing clothes, using curtains, food and drink according to season climatic conditions, no matter they live in their original environment or they move to a new one (Fuller and Bulkeley 2013). A field study in the South Australia demonstrated that residents of low to middle income housing tried to adjust to their environment through less expensive means such as changing their clothes, using fans and curtains, operating openings (Soebarto and Bennetts 2014), and use traditional heating and cooling systems and do not use the air conditioning because they find no need to use it (Fuller and Bulkeley 2013). While other people passive actions were just operating the openings as shown in Japan study (Rijal et al. 2013) and were using openings to control their environment during the summer, and more outfits to provide insulation during winter in other research in Taiwan for the elderly people (Hwang and Chen 2010). A study in China during the cold period showed that the residents who used the personal control presented lower neutral operative temperature than those residents without control (Luo et al. 2014; Cao et al. 2014). Another study in the west part of Athens during the summer period showed that in the cases of the natural ventilated buildings, the thermal sensations are well correlated with the operative temperature compared to the cases of the mechanical ventilated buildings (Sakka et al. 2010). Additionally, the rural residents have greater cold tolerance than city residents (Han et al. 2009), while the hot and humid Caribbean climate in Old Havana caused the residents preferring for higher air velocities during summer (Tablada et al. 2009). A significant result of a sleeping environment research demonstrated that the summer indoor temperatures during sleep could be higher than those prescribed by ASHRAE 55 and that would help in minimizing the energy consumption (Lan et al. 2014). Studies compared the thermal comfort between the traditional and modern residential buildings in each of Turkey, India, Indonesia, and Cameroon found that the traditional dwellings of the residents feel more comfortable than the residents of the modern houses (Manioğlu and Yılmaz 2008; Dili et al. 2010; Sari et al. 2013; Nematchoua et al. 2014). Indoor thermal comfort in the houses has been viewed as being highly dependent on the occupants and the way they see the environment.

③ Museums Indoor Thermal Requirements

Many cultural heritage buildings with high historical values are reused as museums or galleries after retrofitting. Despite that were built for other purposes in their own time, so there is a dual task for the preservation and retrofit operations of the cultural heritage buildings that contain historical collections. The heritage buildings need annual treatment and preventive conservation. Special attention should be placed on the properties of the stone and materials used to best aid in the renovation and conservation of the historic buildings (Arroyo and Rosario 2013; Rosario and Arroyo 2013). Relative humidity and temperature are the two most important environmental parameters in controlling the environmental conditions in museums, in which several collections of significant cultural value are housed (Pavlogeorgatos 2003; Camuffo 1998). Consequently, the need is constant for concurrent monitoring for the preservation and preventive conservation of the building and the collections (Corgnati et al. 2009; Camuffo et al. 2002; La Gennusa et al. 2005).

In the last century, international thermal requirements have been determined to preserve collections in museums, the categories of collections materials are classified and divided into groups based upon their sensitivity to the daily relative humidity fluctuations (Royal Ontario Museum 1979). After much research, guidelines have been established for museum management to include constant indoor climate, with some disagreement on the bounds of temperature and relative humidity which are proper for the maintenance of historical materials whether these materials comprise the building fabric or a museum collection. Erhardt and Mecklenburg's study introduced the effects of the extremes of low (below 30% RH) and high (above 70% RH) relative humidity on the degradation processes, mold growth, and materials fragility in the field of the preservation of museums (Erhardt and Mecklenburg 1994). The museum collections' community has determined specific controls on the indoor climate in a historic building or museum to be within 21.1 ± 2 °C and 50% ± 5% RH year-round. Properly managing the museums' climate via heating, ventilation, and air-conditioning systems is economically difficult and consumes energy, especially when controlling the relative humidity year-round. The Smithsonian Institution system, in the USA, has the highest energy cost per square foot for maintenance of the museums for all galleries because of the tighter environmental controls year-round (Camuffo et al. 2001; Mecklenburg et al. 2004). Studies have been performed in climate management and energy efficiency considering the seasonal change in temperature. The data analysis was separated into seasons which required heating and cooling without the exclusion of the more temperate season, in which there is often a need for both heating and cooling (Erhardt et al. 1995). Recent research of energy consumption recommended passive architecture techniques and the use of solar energy to drive cooling cycles for space conditioning of most buildings in Southern Europe (Tsoutsos et al. 2009; Farmaki et al. 2016). Some research recommended isolation of the building elements more than controlling the climate of the museum spaces. The passive ventilation and using the thermal type of construction elements, especially windows, could be the key to reducing the envelope degradation of the building (Balocco and Grazzini 2009; Efficiency Energy for EU Cultural Heritage 2013).

2.2.2 *Indoor Thermal Neutrality*

According to early thermal studies, Victor and Aladar Olgyay found the climatic conditions in which people are in bioclimatically comfortable conditions consist of: relative humidity between 30 and 65%, temperature between 21 and 27.5 °C and wind speed up to 5 m/s (Olgyay and Olgyay 1963). The recommended thermal comfort range of ASHRAE standard is 23 °C < Temp < 26 °C and 30% < Rh < 60% and for ISO 7730 standard is −0.5 < PMV < +0.5, PPD < 10% and 30% < Rh < 60%. There is a difference between summer and winter comfort zones due to seasonal acclimatization relating to climate, clothing, activity, and metabolismic changes. A field study for both air-conditioned and natural ventilated offices in Thailand (Busch

1990, 1992) found that for air-conditioned offices, the neutral effective temperature was 24.7 °C, while for natural ventilated offices, it was 27.4 °C. The neutral temperatures were lower: 24.4 °C for air-conditioned buildings and 22.8 °C for the natural ventilated buildings due to probably the lower value of clothing thermal resistance (clo). For the natural ventilated building sample, the upper boundary of the comfort zone was found to be approximately 31 °C which is significantly higher than the ASHRAE comfort standard 55–92 value. Heidari and Sharples found that the neutral temperature in the naturally ventilated buildings in Ilam, Iran, during the hot summer in the short-term study was 28.4 °C, and 26.7 °C for the long-term study. The neutral temperature during the cold winter in the short-term study was 20.8 and 21.2 °C for the long-term study. Iranian people in naturally ventilated buildings in Ilam were comfortable at indoor higher temperatures than recommended by standards (Heidari and Sharples 2002). A thermal study in Ghadames, Libya, yielded that the occupants were comfortable at temperatures to 35.6 °C in traditional dwellings compared to 30 °C in the air-conditioned dwellings in summer (Ealiwa et al. 1999).

A limited thermal sensation survey in the East Mediterranean countries for our case study (moderate climate/hot period) is also shown in Table 2.2. In a field study made in Lefkosia, Cyprus, Lapithis found that an average 19.5–29 °C is an acceptable temperature and an average of 20–75% is the acceptable range of relative humidity (Lapithis 2015). It is concluded that the passive systems that are most suited for Cyprus, as well as using of natural ventilation and ceiling fans, and 5% north wall openings are sufficient for cross-ventilation during summer nights with the permanent external shading devices and vegetation (Lapithis 2015). According to Sakka's study, the residents comfort zone was between 27 and 28 °C during the summer period in the west part of Athens, where half of his study participants were dissatisfied from the indoor thermal conditions and want to have more air movement (Sakka et al. 2010). A study in Haifa, Palestine, yielded a neutral temperature of 19.5 °C in winter and 26 °C in summer for the dwellings of passive design conditions, and 21.5 °C in winter and 23 °C in summer for the air-conditioned dwellings. It also highlighted the role of local climate, expectations, and available control in thermal adaptation in actual settings (Becker and Paciuk 2009). Another study in the dwellings in the Gaza Strip, Palestine, yielded that the dwellings do not provide comfort for the occupants in both hot and cold periods, where only 17.83% of the respondents felt comfort in their houses, and the respondents' thermal sensation was hot in the summer and cold in the winter. Gaza dwellings' study recommended to activate the energy-saving strategies in the design and to improve thermal performance and envelope of the dwellings (Zogrob 2014).

ASHRAE suggested an optional approach for defining acceptable thermal conditions using an equation for the thermal neutrality (T_n), which resulted from more than 21,000 measurements occurred around the world, mostly in office structures (De Dear et al. 1998; ASHRAE, ANSI 2010). The thermal neutrality is the temperature average for a large sample of people when the individuals feel neither cold nor hot. The thermal neutrality is affected by the climate which the individuals are used to, based on physiological acclimatization and habits, and related to the outdoor mean temperature (T_m). The monthly mean minimum for the coldest month (T_{min}) and

Table 2.2 Overview of field experiments showing the differences of neutral temperature of different climate zones and different ventilation methods

Reference	Location	Time of year	Subjects	Results
Schiller (1990)	San Francisco Bay Area, USA	Winter and summer 1987	304 subjects (187 females, 117 males) in 10 office buildings (2342 visits)	In winter, the measured neutral temperature was 22.0 °C, versus 24.4 °C predicted by PMV. In summer, the measured neutral temperature was 22.6 °C, versus 25.0 °C predicted by PMV
Fan et al. (1993)	Wuxi, China	All year-round	10 students (5 males, 5 females), in residential buildings and a school	People prefer different thermal conditions during long-term exposure without space heating or cooling than based on thermal comfort standards. Local young people accepted operative temperatures of 10–12 °C in winter
Ealiwa et al. (1999)	Ghadames, Libya	Summer 1997–1998	Survey in the naturally and mechanically ventilated houses	PMV model failed to predict comfort temperatures adequately. The comfort temperature was 35.6 °C in the traditional buildings and 30 °C in the air-conditioned buildings
Nicol et al. (1999)	Karachi, Multan, Saidu Sharif Islamabad, Peshawar, and Quetta, in Pakistan	(1) Longitudinal in summer and winter/(2) transverse with monthly surveys over a year	Both residential and commercial buildings. (1): (36 subjects, 4927 questionnaires)/(2): (846 subjects, 7112 datasets)	PMV tended to overestimate the impact of high indoor temperatures especially in summer time conditions, over emphasizing the need for air- conditioning. There was generally little discomfort at indoor globe temperatures between 20 and 30 °C

(continued)

Table 2.2 (continued)

Reference	Location	Time of year	Subjects	Results
Heidari and Sharples (2002)	Ilam, Iran	Summer and winter 1998, and whole year 1999	Questionnaires in naturally ventilated buildings. Hot summer (513), Cold winter (378), whole year(3819)	Summer neutral temperature in the short-term study was 28.4 °C, and 26.7 °C for the long-term study. The winter neutral temperature in the short-term study was 20.8 °C, and 21.2 °C for the long-term study. People in naturally ventilated buildings were comfortable at indoor higher temperatures than recommended by standards
Fato et al. (2004)	Bari, Italy	Summer (1995, 1999), and winter (1996, 2000)	Air-conditioned building in winter, and naturally ventilated in summer/Sample size: 423, 1034, 250 and 133 in 1995, 1996, 1999, and 2000, respectively	Neutral temperatures were 24.4 and 26.3 °C in summer 1995 and 1999, respectively, and 20.7 and 20.6 °C in winter 1996 and 2000, respectively Occupants of naturally ventilated buildings (summer) regarded a 3.3 and 2.1 K band width to be acceptable compared to 3.6 K in air-conditioned buildings (winter)
Yamtraipat et al. (2005)	Thailand (Chiang Mai, Bangkok and Mahasarakham, Prachuabkirikhan)	August 2001	Users of AC buildings in private and public sectors (1520)	The neutral temperature of people with a postgraduate education level was the lowest around 25.3 °C, while that of the other groups (graduate and scholar) was higher at 26 °C. People with higher educational degrees are found to prefer lower indoor temperature compared to the less-educated

(continued)

Table 2.2 (continued)

Reference	Location	Time of year	Subjects	Results
Becker and Paciuk (2009)	Haifa, Palestine	Summer and winter	A field measurement/survey in the passive and air-conditioned dwellings (189 dwellings in winter) and (205 dwellings in summer)	– The local climate, expectations, and available control have an essential role as contextual variables in thermal adaptation in actual settings – Survey results refuted the model's second assumption that symmetrical responses in the negative and positive directions of the scale represent similar comfort levels – The summer neutral temperature was 26 and 23 °C for the naturally ventilated houses, and the air-conditioned houses, respectively/The winter neutral temperature was 19.5 and 21.5 °C for the houses of passive design, and the air-conditioned houses, respectively
Sakka et al. (2010)	West part of Athens	Summer period (May, June, July, August), 2009	Questionnaires/field measurement for eight naturally ventilated and two mechanical ventilated residential buildings	– Half of the respondents were dissatisfied with their thermal environment and wanted more air movement/The thermal sensations in naturally ventilated study cases were well correlated with the operative temperature compared with the cases used the mechanical ventilation. Summer residents' comfort zone ranges from 27 to 28 °C – The doubt of the thermal comfort indices ability to predict the thermal sensation of the residents in the naturally ventilated buildings in the hot period

(continued)

Table 2.2 (continued)

Reference	Location	Time of year	Subjects	Results
Toea and Kubotaa (2013)	Several countries and climatic zones over the world	Winter and summer	Analysis using the statistical meta-analysis of the ASHRAE RP-884 database	– Development of an adaptive thermal comfort equation for naturally ventilated buildings in three climatic zones hot–humid, hot–dry, and moderate climates using ASHRAE RP-884 database
Zogrob (2014)	Gaza, Palestine	Summer and winter	Comparative analysis with ASHRAE standards/field survey (93 houses)	– The dwellings in the Gaza Strip do not provide comfort for the occupants in both hot and cold periods, where only 17.83% of the respondents felt comfort in their houses – The temperature of the air conditioning adjusts in summer between 18 and 26 °C, and the electric heating device is commonly used in winter with 44.1% of the occupants – The respondents' thermal sensation was hot in the summer and cold in the winter. The predicted indoor temperature equation in Gaza was found: $T_{indoor} = 19.113 + 0.334T_{outdoor}$ – The study recommended to activate the energy-saving strategies and to improve thermal performance and envelope of the dwellings

(continued)

Table 2.2 (continued)

Reference	Location	Time of year	Subjects	Results
Lapithis (2015)	Lefkosia, Cyprus	Year-round	The psychometric chart, Olgyay's bioclimatic chart, Humphreys' comfort chart and Szokolays' equation/questionnaire survey for the experimental solar house (ESH) and the contemporary buildings	– The comfort zone is between 19.5 and 29 °C, and the acceptable range of relative humidity is 20–75%. The experimental solar house (ESH) at Lefkosia meets the energy efficiency goals with an annual energy use less than the contemporary house and the traditional one with 121 KWh/m^2 and 243 kWh/m^2, respectively – Passive solar design and systems of modern buildings guide in Cyprus:(Using the natural ventilation and ceiling fans/direct gain system/permanent external shading devices, vegetation/low emissivity double-glazed argon-filled/external insulation on walls and roof with 0.6 and 0.3 W/m^2K, respectively/Interior thermal storage constructed from bricks and concrete/18% is the sufficient south wall openings ratio/sufficient openings ratio of 5% for the north wall to provide summer cross-ventilation at night)

Source Author and Van Hoof (2008)

mean maximum for the warmest month (T_{max}) temperatures can be found as following according to the recommended working method of Auliciems and Szokolay (Auliciems and Szokolay 1997):

$$\text{The mean temperature } T_m = (T_{min} + T_{max})/2 \tag{2.1}$$

Based on de Dear et al. equation, the thermal neutrality can be calculated using T_m value into Eq. 2.2:

$$T_n = 17.8 + 0.31\, T_m \ldots.(\text{de Dear et al. eq}) \tag{2.2}$$

Here, the comfort zone can then be taken as:
$T_n \pm 2.5$ °C (i.e., 5 K wide) for 90% acceptability.
The range can be extended to be:
$T_n \pm 3.5$ °C (i.e., 7 K wide) if 80% acceptability is adequate.
Auliciemes and Humphreys used laboratory and field measurements in which people were thermally investigated under different conditions, to define the temperature at which they feel comfortable, and to examine the thermal neutrality of the human body. The findings of their studies were statistically analyzed by using regression analysis. Humphreys showed that 94% of the neutral temperature is connected to the variation of outdoor mean temperature (Humphreys 1976). Their regression equations for free-running structures are as following;

$$T_n = 11.9 + 0.534\, T_m \ldots.(\text{Humphreys. eq}) \tag{2.3}$$

$$T_n = 17.6 + 0.314\, T_m \ldots.(\text{Auliciemes. eq}) \tag{2.4}$$

Humphreys, Auliciems, and Nicol have found that comfort temperature can be obtained from the mean outdoor temperature, and they have established the equations of the relationship between the comfort temperature and the outdoor temperature (Humphreys 1978; Auliciems 1981, 1983; Auliciems and de Dear 1986; Nicol et al. 1999). The difference between the equations confirms that there is no universal comfort temperature. Each community must have its own perception of the thermal comfort according to its climate, local culture and type of buildings (Bouden and Ghrab 2005). The thermal comfort zones depend on regional climate. A number of studies have been conducted in naturally ventilated buildings in the different climates in order to investigate the thermal sensation of people and find the comfort zone for the studied areas. Tables 2.3 and 2.4 show the differences of neutral temperature of different climate zones and different ventilation methods.

Toea and Kubotaa tried to develop the adaptive thermal comfort equations for naturally ventilated buildings in hot–humid, hot–dry, and moderate climates using ASHRAE RP-884 database (Toea and Kubotaa 2013). For instance, in their study most of the data for moderate climate appear below the daily mean outdoor air temperature of 20 °C, and the regression lines for moderate climate are defined by the following equation:

Table 2.3 List of neutral temperature of subjects in hot–humid climates

Year	1990	1991	1994	1998	1998	1998	2003
Researcher	J. E. Bush	R. J. De Dear, K. G. Leow et al.	R. J. De Dear, M. E. Fountain	T. H. Karyono	W. T. Chan et al.	A. G. Kwok	N. H. Wong et al.
Building	Office	Residential and office	AC office	Office	Office	Classrooms	Classrooms
Location	Bangkok, Thailand	Singapore	Townsville, Australia	Jakarta, Indonesia	Hong Kong	Hawaii, USA	Singapore
Neutral temperature of subjects	24.5 °C (ET) for AC buildings 28.5 °C (ET) for NV buildings	24.2 °C (to) for AC buildings 28.5 °C (to) for NV buildings	24.2 °C (to) in the dry season 24.6 °C (to) in the wet season	26.7 °C (to) for AC buildings	23.5 °C (to) for AC buildings	26.8 °C (to) for AC classrooms 27.4 °C (to) for NVclassrooms	28.8 °C (to) for NVclassrooms

Source Hwang et al. (2006), Yau and Chew (2014)

Table 2.4 Toea and Kubotaa's adaptive equations based on the outdoor air temperature for different climate zones of ASHRAE RP-884 database

Climate	Hot–humid climate		Hot–dry climate		Moderate climate	
Location; Season–Refined database	Bangkok; Jakarta; Brisbane summer; Singapore (1673)		Karachi summer and winter; Multan summer; Peshawar summer; Quetta summer; Saidu, Sharif summer; Athens summer (2776)		Melbourne summer; Peshawar winter; Quetta winter; Saidu Sharif winter; Oxford, summer; San Francisco summer and winter; Liverpool summer and winter (3213)	
Outdoor air temperature	Adaptive equation	R^2	Adaptive equation	R^2	Adaptive equation	R^2
Daily mean, T_{outdm}	$T_{neutop} = 0.57T_{outdm} + 13.8$	0.64	$T_{neutop} = 0.58T_{outdm} + 13.7$	0.59	$T_{neutop} = 0.22T_{outdm} + 18.6$	0.09
Monthly mean, T_{outdm}	$T_{neutop} = 0.53T_{outdm} + 14.5$	0.51	$T_{neutop} = 0.56T_{outdm} + 14.3$	0.6	$T_{neutop} = 0.18T_{outdm} + 19.3$	0.08
Running mean, T_{outdm}	$T_{neutop} = 0.55T_{outdm} + 14.2$	0.54	$T_{neutop} = 0.63T_{outdm} + 12.4$	0.59	$T_{neutop} = 0.33T_{outdm} + 17.4$	0.09
Prevailing mean, T_{outdm}	$T_{neutop} = 0.54T_{outdm} + 14.6$	0.53	$T_{neutop} = 0.6T_{outdm} + 13$	0.58	$T_{neutop} = 0.35T_{outdm} + 17.1$	0.09

Source Toea and Kubotaa (2013)

$$T_{neutop} = 0.22T_{outdm} + 18.6\ldots.\text{(Toea and Kubotaa. eq)} \quad (2.5)$$

where $R^2 = 0.09, p < 0.001$.

The regression line for moderate climate showed a slightly lower regression coefficient than that of ASHRAE. The predicted neutral operative temperature was between 19 and 24.7 °C. The preferred temperature nearly matches the neutral temperature for moderate climate, which is similar to the findings reported by de Dear and Brager (2002).

2.3 Indoor Thermal Environment Assessment

Achieving human thermal comfort with the lowest energy usage is the main objective of the indoor thermal environment assessment of the building. Several field studies based on the model of the adaptive thermal comfort have been made in many countries worldwide with the aim to have a better understanding of thermal comfort to face climate change and the demands for more energy-efficient buildings (Rupp et al.

2015). The buildings differ in their design, heating and cooling systems, and their occupants' factors such as the clothing and the activities (De Dear and Brager 2002; Nicol and Humphreys 2004). The thermal comfort range is directly related to the condition of the indoor environment, the characteristics of users, the building design and its environmental control systems, and outdoor climatic conditions that has a greater impact on the comfort and satisfaction of the indoor environment (Brager and de Dear 2000; Frontczak and Wargocki 2011). According to American Society of Heating, Refrigerating, and Air-Conditioning Engineers (ASHRAE), the six variables of the thermal comfort are related to the ambience physical factors and to the occupant's personal adaptation to the local climate (air temperature, globe temperature, air velocity, air relative humidity, metabolic rate, and clothing insulation). The closing insulation value is significant to humans' survival. Although sometimes fashion will take precedence over comfort, people generally adjust their clothing to take into account the temperature of their surroundings, hence making themselves more comfortable. Metabolic rate is also used to determine the chemical energy transformation into heat and mechanical work through the human activity inside his building.

To assess whether or not an indoor thermal environment is comfortable, normally two approaches are applied: the physical measurements and the surveys involving questionnaires. The environmental measurements of the relevant thermal comfort parameters in the historic buildings usually accompanied with a thermal questionnaire to evaluate thermal sensation and thermal preference (a five or seven-point scale) of participants. The field survey can help to review the activities level and how occupants dress indoors to assist in determining how they adjust to the temperature levels, for reason of becoming more comfortable.

2.3.1 Microclimate Field Measurements

The physical measurements of indoor thermal parameters and comparisons to existing international versions of the thermal comfort standards such as ASHRAE 55 or ISO 7730 or EN15251 make it possible to determine whether the thermal variables are within the comfort range (ASHRAE, ANSI 2004; ISO7730 1993; BSI 2008). The indoor air temperature, air velocity, and relative humidity variation and average can be recorded in the site by the proper equipment for at least three continues days at a height of 1.1 m above the floor (ASHRAE, ANSI 2004). The simultaneous measurements of indoor and outdoor parameters provide a clear image of the environment conditions inside the historic buildings. The operative temperature (OP) can be calculated based on the mean radiant temperature, air temperature, and airflow that would produce the acceptable temperature (ISO 1998), or only based on the air temperature when the airspeed is lower than 0.2 m/s (ASHRAE, ANSI 2004):

$$\mathrm{OP} = t_{\mathrm{o}=}A\, t_{\mathrm{a}} + (1 - A)t_{\mathrm{mr}} \tag{2.6}$$

(ASHRAE, ANSI 2004), where

OP = operative temperature (°C)
A = 0.5 for small relative airspeed (<0.2 m/s)

Or

$$OP = t_o = \frac{\left(t_{mr} + \left(t_a \times \sqrt{10v}\right)\right)}{1 + \sqrt{10v}} \tag{2.7}$$

(ISO 1998), where

t_a = air temperature (°C)
t_{mr} = mean radiant temperature (°C)
v = air velocity (m/s)

To calculate the mean radiant temperature, the black globe temperature (T_{GN}) should be measured or calculated based on the indoor air temperature value using the following formula (de Abreu et al. 2011):

$$t_{mr} = T_{GN} + 2.42_V(T_{GN} - T_a), \tag{2.8}$$

where:
T_{BS} = dry bulb temperature (°C) = T_a = indoor air temperature (°C)
T_{GN} = black globe temperature (°C), V = indoor air velocity (m/s)

$$T_{GN} = 0.456 + 1.033\, T_{BS} \tag{2.9}$$

(de Abreu et al. 2011).

The aim of physical measurements is to quantifying indoor climate and in particular thermal environment and its main parameters in order to calculate different thermal comfort indices, especially the PMV and PPD that used by the ISO 7730 standard (ISO7730 1993). The Predicted Mean Vote (PMV) is a model uses heat balance principles to relate the six factors of thermal comfort. The PMV is a parameter for assessing thermal comfort in an occupied zone, and it predicts the mean value of the votes of a large group of people based on the ASHRAE thermal sensation scale (−3 cold, −2 cool, −1 slightly cool, 0 neutral, +1 slightly warm, +2 warm, +3 hot). The Predicted Percentage of Dissatisfaction (PPD) of the people is related to PMV. The PPD presents the percentage of people who should be dissatisfied with their climate conditions. PMV and PPD have a linear relationship; they can be directly transformed into one another. As PMV changes away from zero to either the positive or negative direction, PPD increase. It is based on the assumption that people voting (−3, −2, +2, +3) on the thermal sensation scale are dissatisfied. The comfort zone is defined within the recommended limits of the PMV (−1, 0, +1), and the ideal range of acceptable thermal environment for general comfort and neutral sensation should be between [$-0.5 < PMV < +0.5$], with PPD value be less than

10% (ASHRAE, ANSI 2010). If the resulting PMV value generated by the model is within the recommended range, the conditions are within the comfort zone.

Fanger's traditional PMV–PPD method for determining human comfort zone based on heat balance theory is used in the international standards of ASHRAE Standard 55-92 and ISO 7730 (Fanger 1972; ASHRAE, ANSI 1992; ISO7730 1994). Several studies have actually revealed that the natural ventilated structures can be comfortable year-round, and they utilize less than fifty percent of the energy of those structures with air conditioning. Additionally, their occupants were found to accept and prefer a considerably wider range of temperatures that fall out of the typical comfort zone specified by ISO 7730 (Tselepidaki et al. 1992). The adaptive PMV–PPD model is regarded as a practical first step that allows designers to adopt a more complex, socially, and environmentally responsive approach. The model of adaptive thermal comfort used in ASHRAE 55 was derived from the ASHRAE RP-884 database (de Dear et al. 1997). Such database contains around 21,000 actual votes from field studies in 160 office buildings from nine countries located on four continents. The adaptive thermal comfort model used in EN 15251 was developed from the European project Smart Controls and Thermal Comfort (SCATs) (McCartney and Nicol 2002). The SCAT database is composed of approximately 5000 subjective thermal responses from field studies in 26 office buildings located in five countries in Europe. The later update of ASHRAE Standard 55-2004 and ISO7730-2005 used the adaptive comfort model (De Dear and Brager 2002; ISO7730 2005; ASHRAE, ANSI 2004). The Chinese and Asian researchers found that ISO and ASHRAE standards are proper for the people of northern Europe and North America and cannot be applied in Asian context (Olesen and Parsons 2002; ISO7730 2005; Li et al. 2012) (Table 2.5).

Providing the acceptable and healthy environment without heating and cooling system is a big challenge in the era of energy efficiency. It could be a step to save the multiple significant values for the inhabited historic buildings and encourage the continuity of their use adaptively in line with their original designs. Three modifications of the PMV which is (PMVe, aPMV, and PMV*) have been introduced after many laboratory and field study to minimize the energy usage and concern the free-running building, as shown in Tables 2.6, 2.7, and 2.8. These studies mentioned several dynamic factors such as the climate, building type, economics, culture, and user behavior. The Chinese thermal assessment standards have a big contribution in this direction, and they developed their local thermal standards tailored to Chinese climatic zones and the dynamic adapting behavior of the Chinese people in the non-air-conditioned buildings. They established their Adaptive Predicted Mean Vote (aPMV) model based on the laboratory and field studies to design naturally ventilated buildings can minimize energy consumption (GB/T 18049-2000 2000; GB 50736-2012 2012; Li et al. 2012).

Table 2.5 International standards of the range of PMV and other thermal parameters

Standards		PMV	PPD	T	RH	V (m/s)	M (met)	F (clo)
ASHRAE 55-2004		−0.5 to 0.5	<10	/	/	<0.2	1–2	<1.5
ISO7730		−0.7 to 0.7	<15	/	/	/	/	/
Chinese standards	Moderate thermal environments–Determination of the PMV and PPD indices and specification of the conditions for thermal comfort (GB/T 18049-2000)	−1 to 1	<27	/	/	/	/	/
	Code for design of heating ventilation and air conditioning for civil buildings (GB 50736-2012)							

Table 2.6 PMVe model for non-air-conditioning environment (Fanger and Toftum 2001)

Model equation	Expectancy factor of rooms without air conditioning in warm climates		
PMVe = PMV*e	Expectancy	Building classification	Expectancy factor
	High	The building with air conditioning is general, and the summer has only a short period of hot time	0.9–1.0
	Middle	The building with air conditioning is partial, and the summer is hot	0.7–0.9
	Low	The building with air conditioning is rarely used, and the climate is hot throughout the year	0.5–0.7

Table 2.7 aPMV model for non-air-conditioning environment (Yao et al. 2009; De Dear and Brager 1998; Chongqing University, Chinese Academy of Architectural Sciences 2012)

Values of the adaptive coefficient λ				Model equation	
Climate zones		Types		$a\text{PMV} = \frac{\text{PMV}}{1+\lambda\,\text{PMV}}$	
		Residential buildings, small department store, lodge, office	Meeting room, Classroom	**Assessment Category for indoor thermal environments in free-running buildings**	
Very cold, cold zones	PMV ≥ 0	0.24	0.21	Category	Assessment index
	PMV < 0	−0.50	−0.29	I (90% satisfactory)	−0.5 ≤ a.PMV ≤ 0.5
Hot summer and cold winter, hot summer and warm winter, and mild zones	PMV ≥ 0	0.21	0.17	II (75% satisfactory)	−1 ≤ aPMV < −0.5 or 0.5 < aPMV ≤ 1
	PMV < 0	−0.49	−0.28	III (Unacceptable)	aPMV < −1 or aPMV > 1

Table 2.8 PMV* model for non-air-conditioning environment (Zhu et al. 2008)

Model equation
$\Delta PMV = PMV - TSV = 6.961 - 0.26t_{op} - 0.262t_{out,d} - 5.104v$ $PMV^* = PMV - \Delta PMV$
Only applicable in the neutrality and neutral partial heat non-air-conditioning environment

2.3.2 Thermal Field Surveys

The concept of psychometrics became common because it considers the factors of the local environment, thermal parameters, and occupants' comfort and satisfaction. The evaluations of climate change problems and thermal comfort using human subjects and evidence are a superior contribution to knowledge, having a longer-lasting value to the research community than theoretical and simulated evaluations and models (de Dear et al. 2013). In the field survey, occupants are asked one or more questions, which can range from a single, direct question about whether or not the occupant is comfortable, to an extensive questionnaire (McIntyre 1980). Questionnaires may include background information, subjective measures of occupant clothing, level of activity, occupant's sensation and satisfaction votes, and different thermal environment control means, such as the windows, fans, shading devices, etc. Thermal sensation and satisfaction surveys use the numerical representation and Likert rating scale of five and seven-point voting for the six parameters of thermal comfort and the overall satisfaction. Based on ASHRAE Thermal Sensation Scale (2004), the acceptable thermal parameter should have 80% of occupants vote for the central three categories (−1, 0, +1) for the occupants' votes (ASHRAE, ANSI 2004). Thermal sensation vote (TSV), the humidity predicted vote (HPV), and the draft perception vote (DPV) are the main questions to identify the people thermal perception. The subjective voting scale for thermal preferences and acceptability can strengthen the evaluation and give a clear image about the people adapt to their environment and the energy consumption to feel comfortable.

Thermal sensations are different among people, even though they are in the same environment. To reduce the individual differences, de Dear suggested that the bins' mean thermal sensation votes, rather than the individual actual votes, could be used in the analysis (De Dear et al. 1998). The sensitivity of the respondents' thermal sensation to the operative temperature could be evaluated by examining actual mean vote response which forms the basis of research into thermal neutral temperature. Simple linear regression between thermal sensation and operative temperature could be performed and analyzed to specify the strength of the relationship between them in the historic buildings as an organized method to predict the subjective comfort which results from a given temperature. The gradient of the regression model is related to the sensitivity of thermal sensation to the operative temperature (De Dear et al. 1998). The slope of the regression line presents how quickly people felt too warm or too cool as the temperature deviated from the optimum. The estimated neutrality condition is derived by solving the regression equation for zero (neutral), and the comfort band is yielded coinciding with (−1 and +1) sensation votes.

Many field studies have been bringing valuable information from the people conducting their activities in their environments, including research into the person–environment relationship and the factors that affect thermal comfort in the built environment. Studies are still numerous in controlled environments and analyze issues individually. In some situations, the thermal comfort cannot be fully explained by the classical six variables (two human and four environmental). There are a number of other factors that influence the sensation of thermal comfort, like cultural and behavioral aspects, age, gender, space layout, possibility of control over the environment, user's thermal history, and individual preferences. Static and homogeneous environments leading to thermal monotony, an expensive solution, previously preferred, are giving way to dynamic environments, in which wider ranges of indoor temperature are preferred and the natural ventilation is desired. The use of personalized conditioning systems is probably the best ways to increase user acceptability with the thermal environment, especially in the existing and historic buildings and where the interventions are very limited, need high costs.

References

Agnew, N., & Demas, M. (2015). *Principles for the conservation of heritage sites in China*. Beijing: ICOMOS China and the State Administration of Cultural Heritage.

Allen Consulting Group. (2005). *Valuing the priceless: The value of historic heritage in Australia*. s.l.: s.n.

Antonio, M.-M., Isabel, T.-A., Cho, S., & Vivancos, J.-L. (2016). Energy efficiency and thermal comfort in historic buildings: A review. *Renewable and Sustainable Energy Reviews, 61,* 70–85.

Arroyo, F., & Rosario, V.-S. (2013). The church of Saint Martin (Trujillo, Spain): Study of the stone degradation. *Journal of Cultural Heritage, 14*(3), e109–e112.

ASHRAE, ANSI. (1992). *Standard 55-1992, Thermal environmental conditions for human occupancy.* Atlanta, GA: American Society of Heating, Refrigerating and Air-Conditioning Engineers.

ASHRAE, ANSI. (2004). *Standard 55-2004, Thermal environmental conditions for human occupancy.* Atlanta, GA: American Society of Heating, Refrigerating and Air-Conditioning Engineering.

ASHRAE, ANSI. (2010). *ANSI/ASHRAE 55-2010 thermal environmental conditions for human occupancy.* Atlanta, GA: American Society of Heating, Refrigerating and Air-Conditioning Engineering.

Atsonios, I., et al. (2013). *Large scale measurements campaign to assess the thermal behaviour of an 18th century historic building in Athens*. European Academy of Bolzano (EURAC).

Auliciems, A. (1981). Towards a psychophisiological model of thermal perception. *International Journal of Biometeorology, 13,* 147–162.

Auliciems, A. (1983). *Psycho-physiological criteria for global zones of building design*. Stuttgart, Hohenheim: s.n.

Auliciems, A., & de Dear, R. (1986). Air conditioning in Australia: Human thermal factors. *Architectural Science Review, 29,* 67–75.

Auliciems, A., & Szokolay, S. V. (1997). *Thermal comfort*. s.l: PLEA.

Balocco, C., & Grazzini, G. (2009). Numerical simulation of ancient natural ventilation systems of historical buildings. A case study in Palermo. *Journal of Cultural Heritage, 10*(2), 313–318.

Bank, H. S. D. (1998). *Canadian Centre for Occupational Health and Safety*. s.l.: s.n.
Barker, P., et al. (2002). *Principles for the conservation of heritage sites in China*. s.l: The Getty Conservation Institute.
Becker, R., & Paciuk, M. (2009). Thermal comfort in residential buildings—Failure to predict by Standard model. *Building and Environment, 44*, 948–960.
Bouden, C., & Ghrab, N. (2005). An adaptive thermal comfort model for the Tunisian context: A field study results. *Energy Build, 37*(9), 952–963.
Brager, G., & de Dear, R. (2000). *A standard for natural ventilation*. s.l.: s.n.
BSI. (2008). *BS EN 15251:2007, Indoor environmental input parameters for design and assessment of energy performance of buildings addressing indoor air quality, thermal environment, lighting and acoustics*. London: British Standards Institute.
Busch, J. (1990). Thermal responses to the Thai office environment. *ASHRAE Transactions, 96*, 859–872.
Busch, J. (1992). A tale of two populations: Thermal comfort in air-conditioned and naturally ventilated offices in Thailand. *Energy and Buildings, 18*(3–4), 235–249.
Camuffo, D. (1998). *Microclimate for cultural heritage*. Amsterdam: Elsevier.
Camuffo, D., Bernardi, A., Sturaro, G., & Valentino, A. (2002). The microclimate inside the Pollaiolo and Botticelli rooms in the Uffizi Gallery, Florence. *Journal of Cultural Heritage, 3*(2), 155–161.
Camuffo, D., et al. (2001). Environmental monitoring in four European museums. *Atmospheric Environment, 1*(35 Supplement), S127–S140.
Cao, B., Zhu, Y., Li, M., & Ouyang, Q. (2014). Individual and district heating: A comparison of residential heating modes with an analysis of adaptive thermal comfort. *Energy and Buildings, 78*, 17–24.
Chongqing University, Chinese Academy of Architectural Sciences. (2012). *Evaluation standard for indoor thermal environment in civil buildings GB/T 50785-2012* (pp. 11–12). Beijing: China Construction Industry Press. (In Chinese).
Corgnati, S. P., Fabi, V., & Filippi, M. (2009). A methodology for microclimatic quality evaluation in museums: Application to a temporary exhibit. *Building and Environment, 44*, 1253–1260.
Daghigh, R., Adam, N. M., & Sahari, B. B. (2009). Ventilation parameters and thermal comfort of naturally and mechanically ventilated offices. *Indoor and Built Environment, 18*(2), 113–122.
de Abreu, P. G., et al. (2011). Estimativa da Temperatura de Globo Negro a Partir da Temperatura de Bulbo Seco. *Engenharia na Agricultura/Engineering in Agriculture, 19*(6), 557–563.
de Dear, R., Brager, G., & Cooper, D. (1997). *Developing an adaptive model of thermal comfort and preference*. Final Report ASHRAE RP-884, s.l.
De Dear, R., & Brager, G. S. (1998). Developing an adaptive model of thermal comfort and preference. *Ashrae Transactions, 104*(1), 73–81(9).
De Dear, R. J., & Brager, G. S. (2002). Thermal comfort in naturally ventilated buildings: Revisions to ASHRAE Standard 55. *Energy and Buildings, 34*(6), 549–561.
De Dear, R. J., Brager, G. S., Reardon, J., & Nicol, F. (1998). Developing an adaptive model of thermal comfort and preference/discussion. *ASHRAE Transactions, 104*(1), 145–167.
de Dear, R. J., et al. (2013). Progress in thermal comfort research over the last twenty years. *Indoor Air, 23*, 442–461.
Deuble, M. P., & De Dear, R. J. (2014). Is it hot in here or is it just me? Validating the post-occupancy evaluation. *Intelligent Buildings International, 6*(2), 112–134.
Dili, A., Naseer, M., & Varghese, T. (2010). Thermal comfort study of Kerala traditional residential buildings based on questionnaire survey among occupants of traditional and modern buildings. *Energy and Buildings, 42*(11), 2139–2150.
Drake, S., de Dear, R., Alessi, A., & Deuble, M. (2010). Occupant comfort in naturally ventilated and mixed-mode spaces within air-conditioned offices. *Architectural Science Review, 53*(3), 297–306.
Dumitru, B., & Dumitru, G. M. (2008). 6 Sigma tools for management of defects by steel casted parts. In *Proceedings of the 10th WSEAS International Conference on Mathematical and Computational Methods in Science and Engineering (MACMESE'08)*, s.l.

Ealiwa, M., Taki, A. H., Howarth, A. T., & Seden, M. R. (1999). Field investigation of thermal comfort in both naturally and mechanically ventilated buildings in Ghadames, Libya. *Proceedings of Indoor Air, 99,* 166–171.

Efficiency Energy for EU Cultural Heritage. (2013). *Summary guide for local decision-makers, technical guidance on energy efficient renovation of historic buildings* (Online). Available at http://www.3encult.eu/en/deliverables/Documents/WP3_D3.6_20130527_P21_Technical_guide_EE_options_for_local_governments.pdf. Accessed 16 December 2015.

Erhardt, D., & Mecklenburg, M. (1994). Relative humidity re-examined. *Studies in Conservation, 39*(Supplement 2), 32–38. In *Preprints of the Contributions to the Ottawa Congress, 12–16 September 1994. Preventive Conservation: Practice, Theory and Research).*

Erhardt, D., Mecklenburg, M. F., Tumosa, C. S., & McCormick-Goodhart, M. (1995). Newsletter (Western Association for Art Conservation). *Determination of allowable RH fluctuations*, 19–23.

Fan, Y., Lang, S., & Xu, W. (1993). Field study on acceptable thermal conditions for residential buildings in transition zone of China. In J.J.K Jaakkola, R. Ilmarinen, & O. Seppänen (Eds.), *Proceedings of Indoor Air '93* (Vol. 6, pp. 109–114), Helsinki.

Fanger, P. O. (1972). *Thermal comfort, analysis and application in environmental engineering*. New York: McGraw-Hill.

Fanger, P. O., & Toftum, J. (2001). Thermal comfort in the future-excellence and expectation. In *Conference on Moving Thermal Comfort Standards into the 21st Century*. Oxford Brookes University. s.l., s.n.

Farmaki, M.-E., Kyriafini, T. I., Assael, M. J., & Kakosimos, K. E. (2016). Simulated and measured performance investigation of a bioclimatic summer residence in Greece. *Fresenius Environmental Bulletin, 25*(8), 2803–2812.

Fato, I., Martellotta, F., & Chiancarella, C. (2004). Thermal comfort in the climatic conditions of Southern Italy. *ASHRAE Transactions, 110*(2), 578–593.

Frey, B. S. (1997). The evaluation of cultural heritage: Some critical issues. In: *Economic perspectives on cultural heritage* (pp. 31–49). London: Palgrave Macmillan.

Frontczak, M., & Wargocki, P. (2011). Literature survey on how different factors influence human comfort in indoor environments. *Building and Environment, 46*(4), 922–937.

Fuller, S., & Bulkeley, H. (2013). Changing countries, changing climates: Achieving thermal comfort through adaptation in everyday activities. *Area, 45*(1), 63–69.

GB 50736-2012. (2012). *Code for design of heating ventilation and air conditioning for civil buildings*. Beijing: s.n. (In Chinese).

GB/T 18049-2000. (2000). *Moderate thermal environments—Determination of the PMV and PPD indices and specification of the conditions for thermal comfort*. Beijing: General Administration of Quality Supervision, Inspection and Quarantine. (In Chinese).

Han, J., et al. (2009). A comparative analysis of urban and rural residential thermal comfort under natural ventilation environment. *Energy and Buildings, 41*(2), 139–145.

Hassan, M., Xie, H., & Rahmoun, T. (2018). Field study in the residential buildings in the old city of Tartous in Syria during the summer period. In *Resilient & Responsible Architecture & Urbanism (RRAU) Conference*. Groningen, Netherlands: s.n.

Heidari, S., & Sharples, S. (2002). A comparative analysis of short-term and long-term thermal comfort surveys in Iran. *Energy and Buildings, 34*(6), 607–614.

Humphreys, M. (1978). Outdoor temperatures and comfort indoors. *Building Research and Practice, 6*(2), 92–105.

Humphreys, M. A. (1976). Comfortable indoor temperatures related to the outdoor air temperature. *Building Service Engineer, 44*(5), 27.

Hwang, R.-L., & Chen, C.-P. (2010). Field study on behaviors and adaptation of elderly people and their thermal comfort requirements in residential environments. *Indoor Air, 20*(3), 235–245.

Hwang, R. L., Lin, T. P., & Kuo, N. J. (2006). Field experiments on thermal comfort in campus classrooms in Taiwan. *Energy and Buildings, 38*(1), 53–62.

Indraganti, M., Ooka, R., Rijal, H., & Brager, G. S. (2014). Adaptive model of thermal comfort for offices in hot and humid climates of India. *Building and Environment, 74*, 39–53.

ISO7726. (1998). *International Standard ISO 7726. Ergonomics of the Thermal Environment, Instruments For Measuring Physical Quantities*. Geneva: International Standard Organization (ISO).

ISO7730. (1993). *International Standard ISO 7730: Moderate Thermal Environments: Determination of the PMV and PDD Indices and specification of the conditions for Thermal Comfort*. Geneva, Switzerland: International Standard Organizanition (ISO).

ISO7730. (1994). *Ergonomics of the thermal environment - Analytical determination and interpretation of thermal comfort using calculation of the PMV and PPD indices and local thermal comfort criteria*. Geneva: International Standard Organization (ISO).

ISO7730. (2005). *Ergonomics of the thermal environment—Analytical determination and interpretation of thermal comfort using calculation of the PMV and PPD indices and local thermal comfort criteria*. Geneva: International Standards Organization (ISO).

Kanter, R. M. (2011). Zoom in, zoom out. *Harvard Business Review,* 89(3), 112-6, 131.

Karjalainen, S. (2009). Thermal comfort and use of thermostats in Finnish homes and offices. *Building and Environment, 44*(6), 1237–1245.

Kiesow, G. (1982). *Eine Einführung in die Denkmalpflege*. s.l: Wissenschaftliche Buchgesellschaft.

Kim, J., & de Dear, R. (2012). Impact of different building ventilation modes on occupant expectations of the main IEQ factors. *Building and Environment, 57,* 184–193.

La Gennusa, M., Rizzo, G., Scaccianoce, G., & Nicoletti, F. (2005). Control of indoor environments in heritage buildings: experimental measurements in an old Italian museum and proposal of a methodology. *Journal of Cultural Heritage, 6*(2), 147–155.

Lan, L., Lian, Z., Pan, L., & Ye, Q. (2009). Neurobehavioral approach for evaluation of office workers' productivity: The effects of room temperature. *Building and Environment, 44,* 1578–1588.

Lan, L., et al. (2014). Experimental study on thermal comfort of sleeping people at different air temperatures. *Building and Environment, 73,* 24–31.

Lapithis, P. A. (2015). Passive solar architecture in Cyprus.

Li, B., et al. (2012). The Chinese evaluation standard for the indoor thermal environment in free-running buildings. In *Proceedings of 7th Windsor Conference: The Changing Context of Comfort in an Unpredictable World Cumberland Lodge*. Windsor, London, UK: Network for Comfort and Energy Use in Buildings.

Luo, M., et al. (2014). Can personal control influence human thermal comfort? A field study in residential buildings in China in winter. *Energy and Buildings, 72,* 411–418.

Luo, M., et al. (2015). Evaluating thermal comfort in mixed-mode buildings: A field study in a subtropical climate. *Building and Environment, 88,* 46–54.

Manioğlu, G., & Yılmaz, Z. (2008). Energy efficient design strategies in the hot dry area of Turkey. *Building and Environment, 43*(7), 1301–1309.

Mason, R. (2002). Assessing values in conservation planning: Methodological issues and choices. In: *Assessing the values of cultural heritage* (pp. 5–30). s.l.: s.n.

Mason, R., & de la Torre, M. (2002). *Assessing the values of cultural heritage.* In M. de la Torre (Ed.). Los Angeles: The Getty Conservation Institute.

McCartney, K. J., & Nicol, F. J. (2002). Developing an adaptive control algorithm for Europe. *Energy and Buildings, 34*(6), 623–635.

McIntyre, D. (1980). Design requirements for a comfortable environment. Bioengineering. In: K. Cena & J. A. Clark (Eds.), *Thermal Physiology and Comfort* (pp. 157–168). Amsterdam: Elsevier.

Mecklenburg, M. F., Tumosa, C. S., & Prid, A. (2004). Preserving legacy buildings. *ASHRAE Journal, 46,* 18–23.

Michael, A., Demosthenous, D., & Philokyprou, M. (2017). Natural ventilation for cooling in Mediterranean climate: A case study in vernacular architecture of Cyprus. *Energy and Buildings, 144,* 333–345.

Nematchoua, M. K., Tchinda, R., & Orosa, J. A. (2014). Thermal comfort and energy consumption in modern versus traditional buildings in Cameroon: A questionnaire-based statistical study. *Applied Energy, 114,* 687–699.

Nicol, F., & Humphreys, M. (2007). Maximum temperatures in European office buildings to avoid heat discomfort. *Solar Energy, 81*(3), 295–304.

Nicol, J., Raja, I., Alauddin, A., & Jamy, N. (1999). Climatic variations on comfortable temperature: The Pakistan projects. *Energy and Buildings, 30,* 261–279.

Nicol, J. F., & Humphreys, M. A. (2004). A Stochastic Approach to Thermal Comfort-Occupant Behavior and Energy Use in Buildings. *ASHRAE Transactions, 110*(2), 554–568.

Nypan, T. (2009). *Effects of European Union legislation on the built cultural.* Norway: Riksantikvaren, Directorate for Cultural Heritage.

Olesen, B., & Parsons, K. C. (2002). Introduction to thermal comfort standards and to the proposed new version of EN ISO 7730. *Energy and Buildings, 34*(6), 537–548.

Olgyay, V., & Olgyay, A. (1963). *Design with climate: Bioclimatic approach to architectural regionalism.* New Jersey: Princeton University Press.

Østergård, T., Jensen, R. L., & Maagaard, S. E. (2016). Building simulations supporting decision making in early design—A review. *Renewable and Sustainable Energy Reviews, 61,* 187–201.

Pavlogeorgatos, G. (2003). Environmental parameters in museums. *Building and Environment, 38,* 1457–1462.

Porfyriou, H., & Re, A. (2015). Monitoring the management of World Heritage Sites: The Italian contribution. In *Chinese Stone Conservation International Symposium.* Dazu, China.: s.n.

Rappoport, P. (2015). *Best Practice in Cultural Built Heritage* (Online). Available at http://www.heritage21.com.au/best-practice-in-cultural-built-heritage/#. Accessed 18 November 2017.

Rath & Strong. (2002). *6 Sigma pocket guide.* s.l.: Rath & Strong Management Consultants.

Rijal, H., Honjo, M., Kobayashi, R., & Nakaya, T. (2013). Investigation of comfort temperature, adaptive model and the window-opening behaviour in Japanese houses. *Architectural Science Review, 56*(1), 54–69.

Roaf, S., et al. (2010). Twentieth century standards for thermal comfort: promoting high energy buildings. *Architectural Science Review, 53,* 65–77.

Ronchi, A., & Nypan, T. (2006). *European legislation and cultural heritage.* In: A.M. Ronchi & T.M. Nypan (Eds.). Milan: Deleyva Editore. ISBN 88-88943-05-6.

Rosario, V.-S., & Arroyo, F. (2013). The cathedral of Jerez De La Frontera (Cádiz, Spain): Stone degradation and conservation. *Journal of Cultural Heritage, 14*(3), e113–e116.

Royal Ontario Museum. (1979). *In Search of the Black Box: A Report on the Proceedings of a Workshop on Micro-climates Held at the Royal Ontario Museum, February 1978* (pp. 37–39). Toronto: Royal Ontario Museum.

Rupp, R. F., Vásquez, N. G., & Lamberts, R. (2015). A review of human thermal comfort in the built environment. *Energy and Buildings, 105,* 178–205.

Sakka, A., Wagner, A., & Santamouris, M. (2010). *Thermal comfort and occupant satisfaction in residential buildings—Results of field study in residential buildings in Athens during the summer period.* Windsor, London, UK: Network for Comfort and Energy Use in Buildings.

Sari, L., Harris, D., & Gormley, M. (2013). Indoor thermal assessment of post-tsunami housing in Banda Aceh, Indonesia. *International Journal for Housing Science and Its Applications, 37,* 161–173.

Schiller, G. E. (1990). A comparison of measured and predicted comfort in office buildings. *ASHRAE Trans., 96,* 609–622.

Soebarto, V., & Bennetts, H. (2014). Thermal comfort and occupant responses during summer in a low to middle income housing development in South Australia. *Building and Environment, 75,* 19–29.

Tablada, A., et al. (2009). On natural ventilation and thermal comfort in compact urban environments—The Old Havana case. *Building and Environment, 44*(9), 1943–1958.

Toea, D. H. C., & Kubotaa, T. (2013). Development of an adaptive thermal comfort equation for naturally ventilated buildings in hot–humid climates using ASHRAE RP-884 database. *Frontiers of Architectural Research, 2*(3), 278–291.

Tselepidaki, I., Santamouris, M., Moustris, C., & Poulopoulou, G. (1992). Analysis of the summer discomfort index in Athens, Greece, for cooling purposes. *Energy and Buildings, 18*(1), 51–56.

Tsoutsos, T., et al. (2009). Development of the applications of solar thermal cooling systems in Greece and Cyprus. *Fresenius Environmental Bulletin, 18*(7b), 1367–1380.

UNESCO. (2003). *Convention for the safeguarding of the intangible cultural heritage*. s.l: UNESCO.

UNI EN ISO 7726. (2002). *Ergonomics of the thermal environment—Instruments for measuring physical quantities*. s.l.: s.n.

Van den Beukel, A., & CIB. (1993). *Building pathology: A state of the art report*. Rotterdam, The Netherlands: International Council for Research and Innovation in Building and Construction (CIB), W086.

Van Hoof, J. (2008). Forty years of Fanger's model of thermal comfort: comfort for all? *Indoor Air, 18*, 182–201.

Watt, D. S. (2009). *Building pathology: Principles and practice*. s.l: Wiley.

Yamtraipat, N., Khedari, J., & Hirunlabh, J. (2005). Thermal comfort standards for air conditioned buildings in hot and humid Thailand considering additional factors of acclimatization and education level, *Solar Energy, 78*, 504–517.

Yang, W., & Zhang, G. (2008). Thermal comfort in naturally ventilated and air-conditioned buildings in humid subtropical climate zone in China. *International Journal of Biometeorology, 52*(5), 385–398.

Yao, R., Li, B., & Liu, J. (2009). A theoretical adaptive model of thermal comfort—Adaptive Predicted Mean Vote (aPMV). *Building and Environment, 44*(10), 2089–2096.

Yau, Y. H., & Chew, B. T. (2014). A review on predicted mean vote and adaptive thermal comfort models. *Building Services Engineering Research and Technology, 35*(1), 23–35.

Zhu, Y., et al. (2008). Experimental study on the influence of operating temperature, clothing thermal resistance and season on human thermal sensation in partial thermal environment. *National Academic Collection of HVAC Refrigeration*. (In Chinese).

Zogrob, S. M. (2014). *Evaluation study of residential buildings' comfort in Gaza Strip (Case Study: Khan Yunis' Residential Buildings)* (Master in Architecture thesis). The Islamic University-Gaza, Gaza, Palestine.

Part II
Field Surveys and Conservation Strategies of Coastal Built Heritage

Chapter 3
Regional Protection and Risk Management for the Coastal Heritage Sites in Syria

The impact of climate change will result in catastrophic human consequences if action is not taken to address these concerns. Strategic urban planning can aid with the expected impacts of the climate changes moving forward. According to NASA reports, the formation of warm ocean currents will lead to accelerated melting of the Arctic ice resulting in sea-level rise. This, in turn, affects human populations in the coastal cities for many of the world's regions for the past and current centuries. This chapter describes the reality of environmental research and standards in Syria and the current challenges. The previous environmental research on the heritage of the Syrian coast, the climatic problems, and the threatened coastal heritage sites in Tartous city has been presented. Besides the international evidence about the rising sea-level risk, a survey of 209 respondents about the current situation of the Arwad's cultural heritage and the effects of climate change on the Tartous coast was conducted to identify the problem and record the locals' concern. The site conservation needs and preserving vision were summarized, followed by a descriptive analysis of all collected data and maps were made. A protection strategy for the coastal areas of climate change in Tartous coast, Syria, was presented. Goals are to provide a basis for future planning strategies in accordance with the problem of sea-level rise (SLR) in the coastal cities and to open the discussions concerning new ideas of protection strategies for the coastal cities.

3.1 Overview of the Syrian Coast and Environmental Research

Syria has a varied vital climatic and environmental status which combines the dry and semi-dry characteristics within the dominant Mediterranean climate. This distinguishes her from other countries and led to the difference and variation in the modes and styles of building thousands of years ago, in order to reach the proportional adaptation with the climate and facts of the local environment (Hassan 2006). The

M. Hassan and H. Xie, *Climate Change and Conservation of Coastal Built Heritage*, https://doi.org/10.1007/978-981-13-8672-5_3

climate was one of the most significant natural elements that affect the comfort of human beings and their dwelling places. Additionally, Syria is characterized by geographical and historical elements that made it a cradle for ancient civilizations and a cultural meeting point between the east and west. Syrian coast had the greatest share of the sequence of civilizations for thousands of years, that appeared in the human discovery of the secrets of agriculture and the invention of the first alphabet and musical note in Ugarit, and the appearance of urban development systems represented by the buildings that the civilization had left, such as palaces, ports, castles, towers for watching and protection, which exist till now (Hassan and Rahmoun 2016). The traditional architecture style in the notable old cities of Syrian coast is influenced by the traditional courtyard structures of the Levant because of the imitation, inheritance of experiences, and other influenced factors. While the traditional architecture in rural areas has been presented in pristine condition, which is influenced by the local available building material and locals' experiences. The mountains played a significant role in guiding and directing entrances and access roads to it and see the disappearance of the internal courtyards and the use of a simple cube shape and the house with two floors and houses are separate from each other as a result of its geographical location. The nature surrounding of the site influenced construction methods with a significant impact, so differences availability of materials are found clearly visible on buildings throughout the Mediterranean, and the used construction materials stretch of sandstone in the coast to limestone in the mountains and sometimes soft brick in areas where much silt. The early climatic treatments in vernacular architecture in Syria were born spontaneously in order to adapt the locals with their climate and natural environment and their need for association with nature and co-existence with it. The architectural treatments in Syria such as in worldwide gradually declined after the industrial revolution until the present time as a consequence of different economic, cultural, social, and technical changes, despite the green building calls and modern environmental trends.

3.1.1 The Current Challenges and Efforts of the Coastal Built Heritage

The Syrian heritage sites and buildings faced many challenges at the national and global levels. The global challenges are due to climate changes as shown in Fig. 3.1. The sea level rises in coastal areas. This made worse by a rise in surface temperatures (Nakicenovic et al. 2000; Hassan and Rahmoun 2016; Cassar and Pender 2005). One factor that has greatly affected Syria's heritage sites is the war which only started in 2011. It has caused many deaths of both the humans and the cultural kind. Like most wars, it has brought destruction upon the land. Numerous cultural properties are at the heart of the regional conflicts and war and faced many threats that affect their value. As a result, UNESCO has included 46 properties on the World Heritage

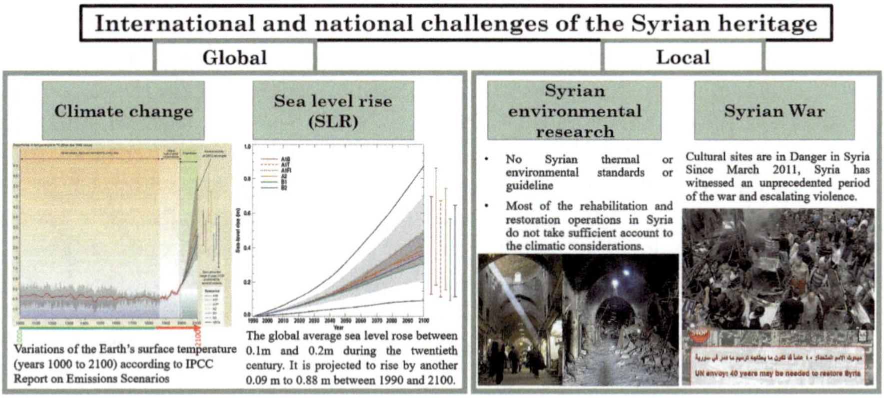

Fig. 3.1 International and national challenges of the Syrian heritage

List in Danger in accordance with Article 11(4) of the convention, to encourage the international community to participate in an effort to save the under risk sites.

There is a need for the restoration and rehabilitation of damaged cultural sites, and efforts to do so have already been made. Syrian Directorate-General of Antiquities and Museums (DGAM) and the competent authorities, in collaboration with the German Archaeological Institute, started the speed work on documenting and digitalizing since 2013. The "Digital Syrian Cultural Heritage Archives Project" aimed to make digital databases of the Syrian heritage and cultural properties which contribute in strengthen the protection process and support the later restoration operations. Then, the DGAM launched a version of the interactive map of the damage of the archeological sites and monuments caused by the armed clashes within the historic sites, the fires, the explosions, illegal secret excavations at archeological sites, etc. (DGAM 2017; Abdulkarim 2013). The universities, local communities, and volunteer networks all over the country participate in protecting Syrian cultural heritage in the times of armed conflict. Syrian government has launched the international calls to action for Syrian and tried to combat the looting of Syrian antiquities and counteract their illicit trade, such as the Unanimously Adopting Resolution 2199 (2015) bans trade in cultural materials illegally removed from Syria, and the cooperation with a lot of countries on the protection of the Syrian cultural heritage. In 2015, DGAM started within the framework of a comprehensive project in cooperation with its partners and under the supervision of UNESCO to document the damaged built heritage and prepare a temporary support study to develop a future plan for the rehabilitation. DGAM in cooperation with the Syrian experts and international companies started conducting the laser surveys using scanners, drones, and scanning technology, as well capturing satellite images providing dependable and high resolution (DGAM 2017; ArchDaily 2017).

However, most of the restoration projects barely considered the relevant climate and academic researches that were made prior to said restoration and rehabilitation activities (Hassan et al. 2017; Al Dakar 1999, 2000a, b, 2005). One can consider the fact that environmental research is rare because not many organizations want to provide funding. Any environmental research could be a great help to the country. The goal is to have heritage sites be able to withstand war or at least minimize the damages that war can bring. Universities' research try to start focusing on monitoring and managing the heritage sites built on so-called safe areas as a step for postwar heritage rehabilitation. As well, Syria is still in need of updating the codes for the improvement of handling solar radiation through roof insulation and for suitable conservation policy for the preservation of historic buildings (Hassan 2006). Special attention should be placed on researching the properties of the stone and materials used to best aid in the renovation and conservation of the historic buildings (Adorni et al. 2010).

① **The Reality of Environmental Standards in Syria**

The international environmental standard versions of Middle East still unsuitable and unable to adapt the local conditions and customs with the lack of cooperation and support from the Arab local organizations, and the absence of sustainable regulations in most Middle Eastern countries regarding energy and water efficiency and indoor environmental sectors. Despite the Arab efforts to have their own rating systems such as Abu Dhabi's Estidama Pearl, Qatar Sustainability Assessment System, and although the emergence of Masdar City, (Komurlu et al. 2015; Attallah et al. 2013), the sustainability still deficient and faced many challenges presented by the lack of metrics, the gap of research and measurement field, monitoring systems, and environmental management and regulations (Komurlu et al. 2014). Despite Syria's concern for the environment and sustainability, the situation of environmental research is still the same unchanged such as another research in other areas if it is not the worst (Haydar and Pediaditi 2010). The scientific research sufferers of the scarcity of production, the weakness in key areas, the near absence in advanced fields such as information technology and measurement field, the low spending on research, and the low of the number of qualified to work in it. In response to the growing international scientific understanding of climate change and anxiety of the implications; Syrian Arab Republic joined the United Nations Framework Convention on Climate Change in 1996, then joined Kyoto Protocol, which entered into force in 2005, and participated in conferences and the negotiations of the parties of climate change, that took place in the working groups, the awareness of the importance of the principles of green architecture and its role in sustainable development, and the risk of the climate change were started in Syria. The Ministry of State for Environmental Affairs set up the first national communication of climate change, including the inventory of greenhouse gas emissions and ways to reduce them, and the preparation of the national strategy to adapt to climate change and entered within the five-year plans of the state in addition to the implementation of some projects, such as the deployment of technologies for the production of biogas and the distribution of solar cookers in the northern region, and the Ministry started the annual National Environmental Research Conference from 2013, to strengthen cooperation with research institutions and the community.

The latest activity was Syrian participation at the twenty-first Conference of the Parties on Climate Change in Paris 2015. On the built environment direction, Ministry of Local Administration in Syria directed toward the observance of environmental conditions within the updating of the building codes for the cities and try to raise awareness and training national cadres by the European Union (MAM project) in several cities. Syrian built environment standards presented by the only Syrian thermal insulation code, which is published by National Syrian Arab Standards and Metrology Organization, are still very limited (SASMO 2015). Despite the existence of the Thermal Standard for Buildings (TSBL) in Lebanon, a neighboring small country to Syria, which established in 2010 and determined the energy-efficient criteria for new building designs (ADEME 2010), Syria still has no similar standards. TSBL yields an envelope compliance certificate based on the climatic regions in Lebanon (coastal region, mid-mountain region, high-mountain region, inland region). It provides a guideline to determine the maximum thermal energy needs for cooling and heating for the whole building, and to minimize the energy consumption of the building envelope by applying the thermal transmittance U-value for building components such as windows and solar shading devices, or for the entire building such as U-building, U-façade, and windows-to-wall ratio (UNDP/GEF and MPWT/DGU 2005).

The current situation of the Syrian housing, changing the climate, increase the low-income people, and building codes and legalization caused disregard of the health and welfare of the inhabitants and their quality of living conditions particularly in the old cities and the historic buildings and communities. The process of promoting and developing research in Syria faces difficulties and obstacles; because of the weak linkages between the research and the development and production sectors of society, the absence of institutional support and the lack of an enabling environment for the development of scientific knowledge, which led to suffocate the local level, in spite of Syria' owning of human resources, are able to stimulate the awakening of knowledge.

② The Built Heritage in the Syrian Coast and the Environmental Research

Syrian studies in the recent decades called for using the components of the vernacular architecture, and locals experience in new design. In the 1980s, Syrian efforts tried to use the local resources and architecture in addition to the regional architectural tradition in building new structures (Aga Khan Award for Architecture 1991). Several local Syrian bodies and universities' projects contributed to documenting the important features and patterns of the traditional and vernacular architecture in Syria and the coastal cities, especially the French and Ottoman periods' styles which are still existed so far. Documentation of the old city of Tartous book recorded the different windows' styles which could help in restoring and establishing a conservation code for the old city of Tartous and help in further research, as shown in Fig. 3.2 (Bashour 2011). A study of the traditional architecture in Latakia from the thermal point of view is called for preserving the advantages of traditional buildings in the new one with recommendations for using the proper location of the openings. The study emphasized the disadvantages of the traditional courtyard dwelling in Syrian

Fig. 3.2 Some recorded traditional windows patterns in the old city of Tartous. *Source* Bashour (2011)

coast due to lack of cross-ventilation in hot humid summer. As well as the advantages of the later opening of the traditional dwellings up to outside which developed later to cover most of the facade overlooking the alley on the upper floor, where the locals didn't prefer openings in the ground floor facades for the privacy and security purposes (Al Dakar 2005). The results yielded the importance of controlling the prevailing and secondary winds by trying to find a wind protection in winter and to provide cross-ventilation and ensure the ventilation in proper speeds to improve the internal atmosphere due to the high relative humidity in summer in Latakia. It also emphasized the importance of balconies' existence as semi-outdoor spaces in suitable locations in the Syrian coast houses and innovated a treatment of the eastern ventilation of the southern balconies by partial closing using a temporary fabric barrier in summer (Al Dakar 2005).

The disappear and decay of the carvings and decoration of the facades, the black color crust in the most stone facades, and the widespread existence of the enhanced biological processes on the stone face and algal and biological soiling growths, as well as the differentiation of the stone decay in the same facade, are the main problems that could be directly recognized by the visual diagnosis. It is so difficult to determine one geological, chemical, or environmental reason of stone decay and deterioration of the historic buildings facades in the Syrian coast. Several factors combined together to cause the decay such as high humid maritime environment and weathering factors, atmospheric pollution, stone durability and porosity, salt effects, and the water damage and moisture ingress into the building, where the combined effects will be greater than the individual one. Damage factors have intermittent nature in terms of their effect and may go through a seasonally or relative period of inactivity. The human intervention and the wrong restoration works cause stone

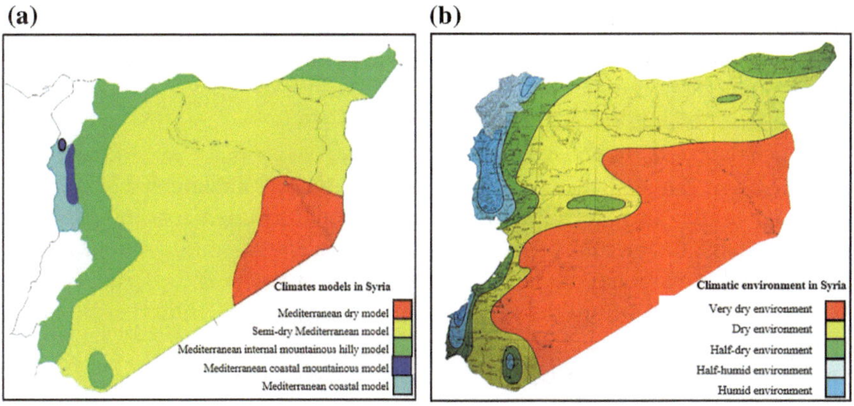

Fig. 3.3 Syrian climatic regions' classification: **a** classification of Adel Abdel-Salam (1973–1990), **b** classification of the general directorate of meteorology

damage such as using inappropriate materials and mortar, using iron wedges, and cleaning and conservation procedures for aesthetic rather than structural purposes (Nezam 2004). Nezam experimented one restoration material on three samples of sandstone from three coastal places; each sample had an original sample from the same origin as a post-experiment reference. Then make an assessment based on observing the changes in color, and the changes in porosity, chemical activity inside the porous, the water absorbing criteria. The sample monitoring for one year demonstrated satisfied results (Nezam 2004).

Several researches divided Syria to climatic regions which can be useful in establishing a thermal standard. The first classification is Adel Abdel-Salam classification (1973–1990) for Syrian climatic regions based on the natural geographic characteristics and locals. The second one is the Climate Atlas of Syria classification, set by the Climate Directorate in Syria and administration 1970 Military Survey (General Directorate of Meteorology in Syria 1977), which depends on the information, experiences, and statistical methods for 15 years as a basis in the preparation and issuance, as shown in Fig. 3.3. The third classification is Hassan classification which combines the previous classifications by division of Syria to four main climatic regions graded into eight local environments based on thermal criteria and the reflection of the climate on the tradition and modern architectural design in Syria (Hassan 2006). Hassan determined the facts and components of each bioclimatic region in Syria, and the positivisms and negativities of each region to conclude the climatic treatment that ought to be presented for each environment. Her research also handled how the existing historic typical designs are suitable for the bioclimatic environment (through analysis and comparison of traditional and modern residential samples of architecture) to secure the human comfort in dwellings through adaptation of architectural designs to the regions' environment where they exist and the use of natural resources (Hassan 2006).

Hassan research studied the differentiation of bioclimatic regions in Syria and its effect on architectural designs using the Olgyay's chart (Olgyay 1953) to determine the human interaction with his climatic environment. For the Syrian coast environment, she used Latakia climate data (longitude: 35–36 E, latitude: 35–36 N, altitude: 2–200 m) in her study to present the whole Syrian coast climate, due to the available Latakia station climatic data of the last decades which are documented in the climate atlas of Syria, and due to the small distance between the Syrian coast cities. According to this study, the months on the Syrian coast are classified into three categories: hot, cold, and transitional months. The hot period of the year in Syrian coast, our case study period, is within the months (6, 7, 8, 9), and the entire period remains over the area of comfort even in September.

According to the human comfort chart of the Syrian coast, Fig. 3.4, the outdoor considerations regarding the building location in terms of dimensions and their relationship with each other, and the treatment of internal atmosphere in summer summarized by the protection from the sun through the roof insulation and secure shadows for buildings. Secure internal ventilation to get into the comfort area, the arrangement of interior and exterior design of buildings and apartments, and choosing the proper size and direction of windows, as well as the appropriate fixed and mobile shading elements create differences in atmospheric pressure. As well as, the critical need for treatment in cold months is to get benefit from the solar radiation, provide protection from winter rains, and provide protection from cold winds. Additionally, to secure the ventilation in hot months, thermal inertia and insulating, and secure the shadows. The general prevailing wind is east to northeast in winter and south to southwest in summer. Due to the coastal mountain range that extended north–south direction and forms a dam in the face of the eastern and western wind, leading to relatively deflected to the current mainstream. In the cold period, especially during the night because of the impact factor (pre-sea), where the rising warm maritime air masses to be replaced by cold air coming from the eastern Interior regions, as shown in Fig. 3.5.

Based on the sun's apparent movement scheme on latitude lines 32–36°N and the scheme of solar radiation on the horizontal and vertical surfaces; in summer, the south and north facades are suitable, and the horizontal surfaces, eastern and western facades, should be protected. While in winter, the advantage of the south facades, and the north, northeastern should be taken, and northwestern facades should be isolated because they are less vulnerable to solar radiation. Therefore, the roofs and the eastern and western facades should be isolated, as shown in Figs. 3.6 and 3.7 (Al Dakar 2000a, b; Hassan 2006). A large percentage of the new and French and late Ottoman periods buildings in Syria have horizontal ceilings of a solid concrete slab with limited thicknesses of 10–15 cm. They do not have self-sufficient thermal insulation, and the additional insulation applied is rarely sufficient if it found. The final floors in the buildings always have some problems as higher temperature degrees at summer and lower temperature degrees in winter than normal floors. Those problems happened because of the thermal exchanges with atmosphere and the radiation with the sky by a big area of the roof (Al Dakar 1999). In aim to treat the final roofs on the Syrian coast by radiation control, Al Dakar examined covering the roof by using aluminum

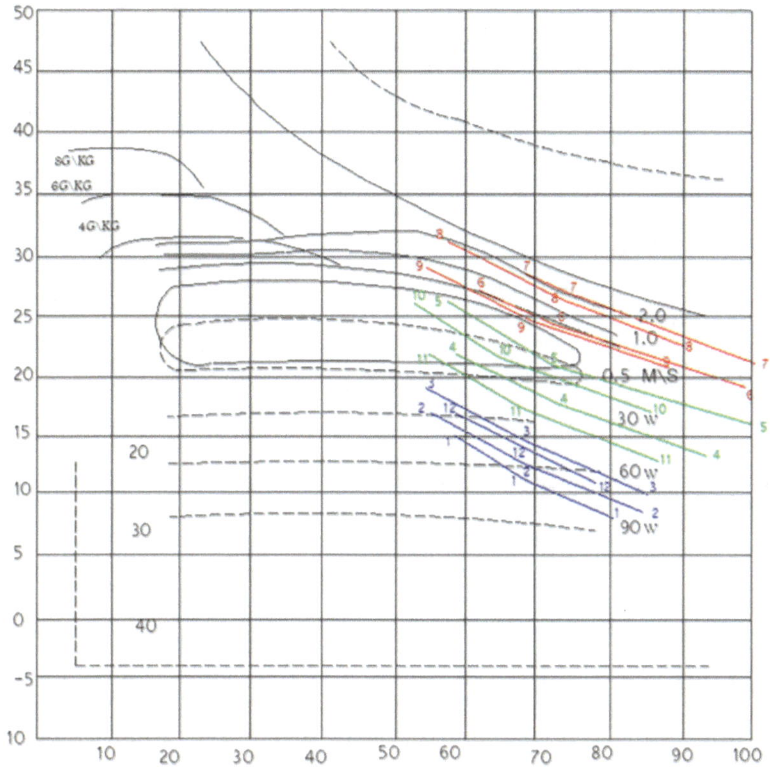

Fig. 3.4 Syrian coast months categories according to the human comfort chart of Lattakia. *Source* Hassan (2006)

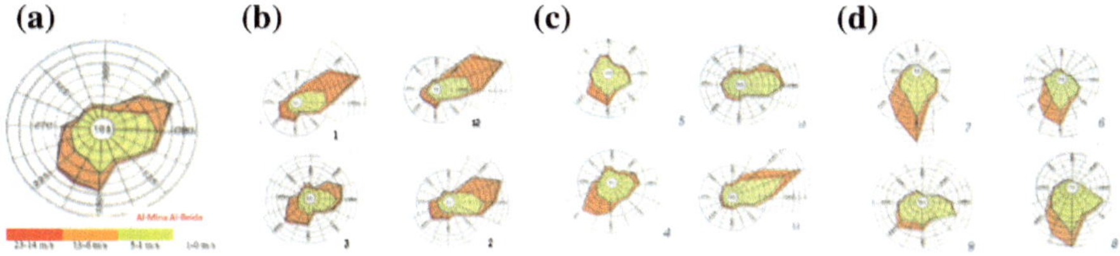

Fig. 3.5 Prevailing wind in Lattakia **a** year-round, **b** in the cold months, **c** in the transitional months, **d** in the hot months. *Source* General Directorate of Meteorology in Syria (1977)

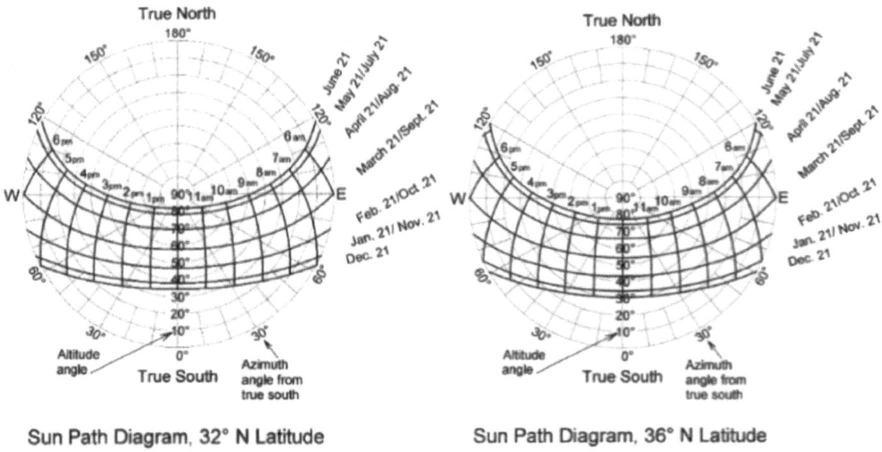

Fig. 3.6 Sun-path diagram, latitude 32°N, 36°N. *Source* http://www.harvestingrainwater.com/sun-shade-harvesting/sun-path-diagrams/

foils above a thick paint of bitumen, and covering the aluminum foils by a paint of calcium hydrate during summer, then cleaning it at the beginning of winter to protect the upper floor of the cold weather in winter and hot weather in summer, where the white raw aluminum is characterized by low radiation exchange and low absorption of solar and earth radiation (Al Dakar 1999). Al Dakar got positive results of these treatments in both winter and summer periods, but some problems had been appeared such as the chemical deterioration of the metallic surface, with the effort and the cost needed to clean it every year at the beginning of winter. He experimented the treatment again with the use of gyps(white plaster) mixed in the paint of calcium hydrate with different percentages on the roof which gave good results of function and protection of aluminum foils. Also to minimize the effort and costs needed for the cleaning of the metallic surface and to make the function of this operation work gradually at the same time with its need (Al Dakar 2000a, b).

There are no Syrian standards or guideline as well as a weak publication and electronic documenting mechanism in each of the government offices, libraries, and preserving academic bodies and associations. Despite the limitations of finding previous Syrian and finding related local research and knowledge about our heritage problems, restoration methods, and treatment, the few previous research with further investigation such as simulation and analysis on the weather data in Syria, the optimal passive design, and efficient conservation outlines could be draft for the Syrian cities based on the climatic zone map (Hassan 2006).

③ The Annual Increase of the Temperature in Tartous Coast

Due to the lack of climatic data in the existence of meteorological station of Arwad and Tartous for several years ago, the available temperature data from 1952 to 2012 of Latakia were used, a nearby coastal city station is located on Lat 35.5 N, Lon 35.8 E

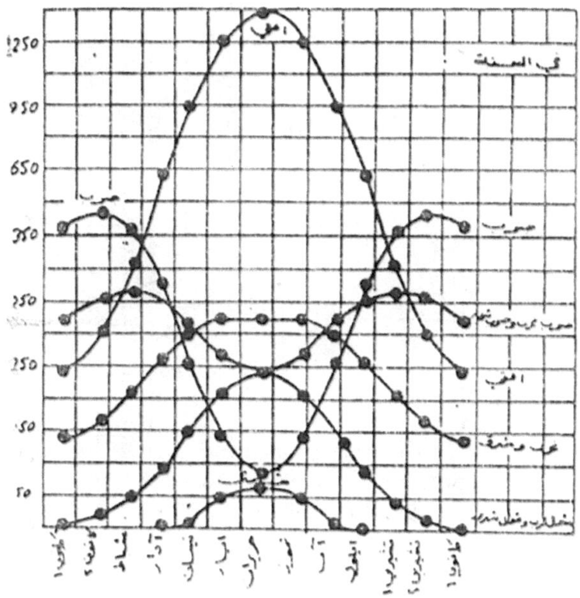

Fig. 3.7 Scheme of solar radiation on the horizontal and vertical surfaces. *Source* Hassan (2006)

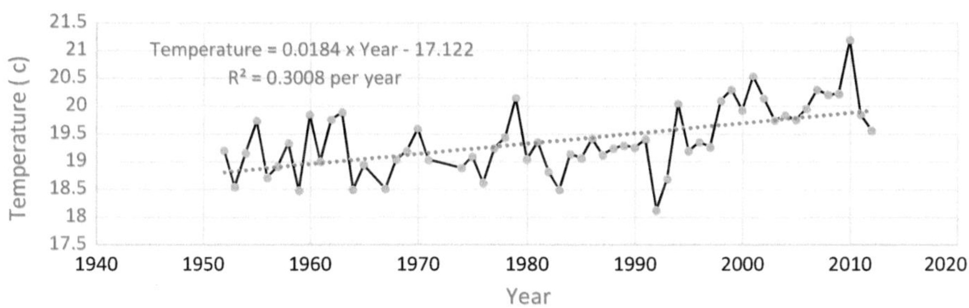

Fig. 3.8 Lattakia temperature history from 1952 to 2012

north of Tartous about 85.85 km (Climate Directorate in Damascus from 1950 to 2012, NASA Goddard Institute for Space Studies 2014). Based on the temperature data of Latakia station, the Climate Directorate in Damascus, and the website of NASA Goddard Institute for Space Studies (GISS Surface Temperature Analysis 2014), and by using the Excel software, the variation of temperature during the past sixty years was found. Figure 3.8 shows that the temperature has increased during the last sixty years with 1.1 °C per 60 years. By finding the equation of the temperature increasing and R^2 value that was small because the scatter values, the expected increase in the temperature for the next one hundred years could be further predicted to be 1.83 °C.

Fig. 3.9 Bad weather and people adapting: **a** boats are used daily to transport people, food and basic life needs, **b** the ship Danny F II sank in 2009, **c** the ruins of the ancient defensive wall, **d** storms in December 2012, **e** the damage and destruction for the western side of the corniche due to storms of January of 2015. *Source* **b** https://en.wikipedia.org/wiki/MV_Danny_F_II#/media/File:MV_Danny_F_II.JPG, **d** Arwad Island port, and the rest photos taken by the Author

④ The Winter Weather and Storms in Tartous Coast

Most storms come from the western and southwesterly directions. North high-speed wind does not cause storms and always cause frost. According to the survey and the interviews with the local people and employees of the coast guard and Tartous navy, the island recently faced three great storms in each of 2009, 2012, and 2015, with a wind speed reached 150 km/h. These storms are the cause for closing the port for many times, the delinquency and sinking of ships laden with many kind goods, livestock, and passenger, and the damage and destruction for the western side of the corniche, such as the school and the concrete floor of the corniche; see Fig. 3.9.

In December 17, 2009, Danny F II was a cargo ship for a livestock, the ship capsized and sank in a storm off Syria 20 km, while it was sailing from Montevideo, Uruguay, to Tartous, Syria, carrying 83 passengers and crew, and about 30,000 sheep and cattle, the passengers are dead, and all the livestock are perished. Rescue efforts were hampered by poor weather conditions, strong winds, severe rain, high waves, and floating carcasses from the ship's livestock cargo. In December 2012, the wind speed reached 150 km/h, two ships aground one had a cargo of cement, and the other one had chemicals cargo to the north shore of Tartous, one of them still in the same place because of the extracted costs. In January 2015, the wind speed of the storm that is called Zena reached 120 km/h and the high waves reached to 10 m, it continued for a period of 7 days, then another storm after one month has followed it in February for 6 days, the first storm was very strong and led to the destruction for some areas in the western side of Arwad Island's Corniche, the fence of Arwad school has fallen, and the concrete floor near the school was damaged.

3.1.2 *The Threatened Coastal Heritage Site in Tartous of Climate Change*

The Syrian coastal zone geographically limited with narrow coastal plain and high population density concentrated in the coastal cities about six times more than the rest of the areas within the region. Total Syrian seafront is not exceeding 183 km length, against the total state area of 185,180 km^2, a ratio of 1012 km^2 for each 1 km seafront, more than five times the Mediterranean states' average of 192 km^2 per linear km of the coastline (UNEP/MAP-METAP SMAP III Project 2009). It also constitutes 2.5% of Syria space, but it includes 11% of the number of Syrians. The strategic importance of the Syrian coast besides the significant specifications and support for the national economy as regards trade and tourism and agricultural activity, including the attributes of the newly discovered oil, emphasizes the importance of protection of the human history, populations, and activities from the risk of SLR (Rahmoun et al. 2016). The book regionally focuses on the vital and future growth area of Tartous city and its coastal heritage and urban sites threatened of SLR problem, which is an important area for starting to take protection measures with the possibility of expanding the model later to different areas of the Syrian coast (Faour et al. 2013). Taking into account the most optimistic scenario of rising sea levels in the northeastern part of the Mediterranean Sea basin, as well as the most pessimistic scenario, the water level rise of 0.5–0.75 m in 2100 (Church et al. 2013). The coastal heritage and urban sites threatened by the climate changes problems, SLR problem, and the increase in temperature in Tartous city presented by the historic Arwad Island, Tartous old city, historical Amrit and its sectors, and Tartous city seafront, as shown in Fig. 3.10. Amrit site was registered as a national heritage, while the national cultural heritage sites of Tartous old city and historical Arwad Island are submitted on the world heritage tentative list since 1999 that stopped registration after 2011 due to Syrian war where Syrian heritage sites and buildings have been classified as danger since 2013.

- Arwad Island, the only inhabited island opposite the Syrian coast, was founded by the Phoenicians as a marine trade base for them, and they built a defensive wall of huge stone square form for protection from high waves and attack enemy ships. As well as the remains of Phoenician defense wall and submerged underwater antiquities, Arwad Island contains the old rock ruins and caves carved into the rocks in the western facade of the island, which back to the mother rock that Arwad Island was built on in the ancient time, archeological buildings represented by the citadel, tower and the remains of the Ottoman bath, and the traditional residential buildings belong to various periods.
- After Phoenicians had founded Arwad Island, they moved to Tartous coast, where they found good soil for agriculture and established the Kingdom of Amrit in the south of the current city near a small river known as Amrit River as a site for their activities, sports, and cemeteries for their people. Historic Amrit site has the remains of the temple in the middle of a large square court defined by a colonnade, and Hellenistic sport stadium with the residential area, funerary monuments, tombs

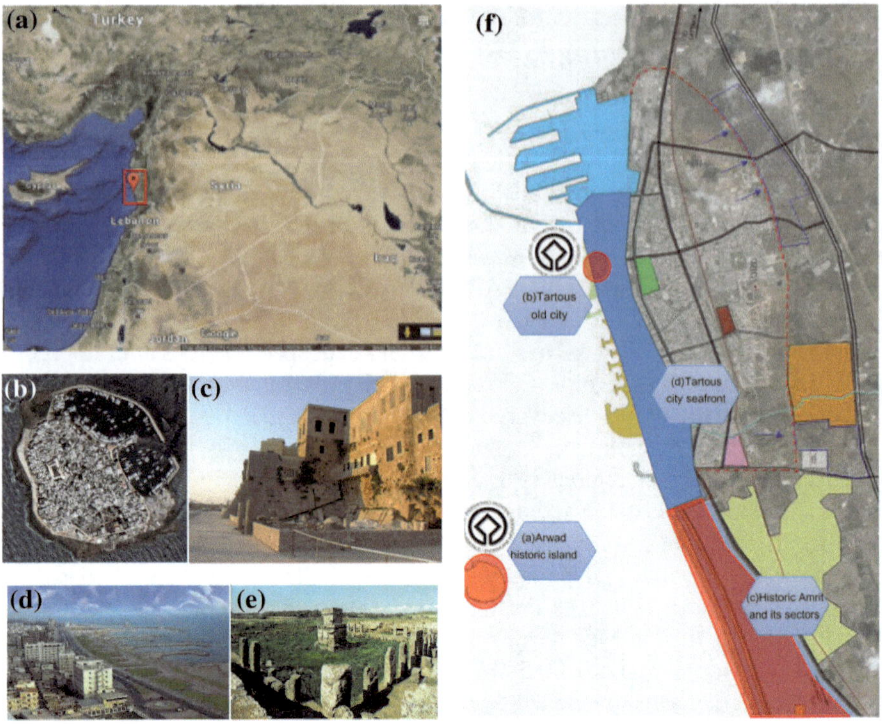

Fig. 3.10 Case study site: **a** geographical location of Syria, **b** historic Arwad Island, **c** Tartous old city, **d** historic Amrit and its sectors, **e** Tartous city seafront, **f** Tartous stone heritage and urban sites are threatened by SLR. *Source* Google Earth, Author

and stone sarcophagi dating back to the third century B. C. Amrit site is divided into sectors which are offered for tourism investment, taking into account the relics.

- Despite the first Phoenician presence in the city of Tartous, but the basic core for the emergence of the old city of Tartous current form dates back to the Knights Templar, when the Crusaders arrived in the Levant, they quickly tightened their control on it, because of its important strategic nautical location, off the coast of Cyprus. Tartous became one of the main settlements of the Knights Templar who built the citadel in the seafront with double concentric walls and a double moat, and the most important Knights Templar buildings presented by Banqueting Hall and the Chapel, also they constructed a lady Tartous Cathedral, Tartous Museum now, in the same location, which is believed to contain the ancient temple of the Virgin Mary. The dwellings were later built within the citadel and the remains of its walls, and the later Ottoman additions such as the galleries along the walls, are still clearly visible.

3.1.3 *Zoom Out Approach*

The field work used Zoom Out and Zoom In approaches for monitoring and preserving processing as modes of strategic thinking for advanced leadership from Harvard Business School (Kanter 2011). These modes are to obtain the virtues of both perspectives, and to create the ability of moving across a continuum of perspectives and Top-Down and Bottom-Up Processing, to face the problems of climate change, and down to find solution for local people needs with a whole image perception. Zoom Out approach used the top-down monitoring and preserving processing, as a stepwise approach, controlled from the high levels of heritage site management and planning, where the regional and urban planning play the role of putting strategies and making decisions starting from wide scope down to smaller segments with details gradually added. The triangulation is concurrently used to compare information from different independent sources, gather the facts and data to evaluate the area, and find out the outlines for drawing together the level of intervention, the proposed scenario, and strategy for protecting Syrian coastal heritage built on the evidences base and systematic review of relevant literature, current situation, and possibilities. The approach procedures:

① **Data Collection and Problem Definition**

This approach applied a series of data collection with the goal of discovering and finding out the solution keys and the resources available for the Syrian coast. Then, a review of many theories and strategies, and collecting the relevant information from the available historical, topographical, and current related maps and the related published literature of different scientific research work.

(1) **Questionnaire Survey**

A total 209 respondents of the Arwadian people were randomly selected for this survey. The survey provided background information on the current situation of the Arwad's cultural heritage and the effects of climate change on the Tartous coast in Syria from the point of view of the island locals, the people who concern on, the stakeholders, and decision-makers. Then Excel spreadsheets and the SPSS 18 software were used to statistically find out the needs and locally strengthen the theoretical background and identify the problems based on the people participation. The content of the survey is shown in Fig. 3.11 and Appendix A.

② **Descriptive Analysis**

After defining the research problem, determine the threatened stone heritage sites of the coast and identify the important data. An elicitation the successful attempts from the related global published papers and cases to use it and reach to the goals. Hence, in order to apply the new scenario of protecting Syrian coastal cities, a descriptive analysis of the spatial coastal current and future situation includes analysis of marine topographic survey schemes in parallel with the analysis of the nature and needs of current communities through three stages:

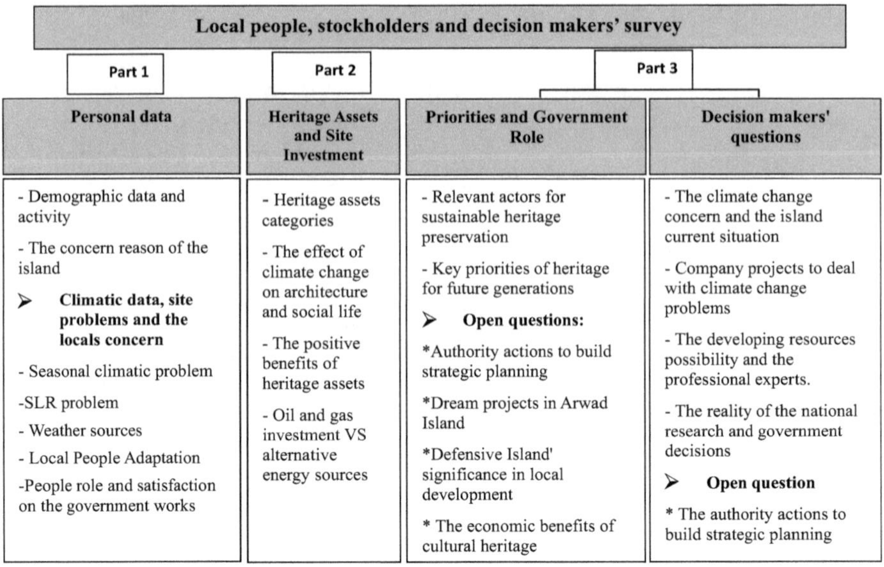

Fig. 3.11 Content of the local people, stockholders and decision-makers' survey

(1) The Possibilities of the Continental Shelf

This analysis and study give us a clear image of the shape of the proposed defenses (a series of islands with continued growth) and places most convenient for its establishment, based on the study of marine depths of the continental shelf, as well as the appropriate directions for future growth.

(2) The Possibilities Within the Territorial Waters and the State Plan for Investment

This analysis specifies the sources and possibilities within the territorial waters, the state plan for investment and the nature of the advantage of defenses (artificial islands viable to connect in the future), in order to develop sustainable development plans. And suggest the nature of the proposed human activity in the future plans, between the shoreline and the defensive line.

(3) The Shoreline Development

This stage includes a study of the emergence and development of the city. Then, the analysis of the current status of the endangered slice of the city and development projects in it.

③ Determine the Procedures and Strategy

After analyzing the collected data and maps and the current situation, the procedures and steps for the development of future plans were determined, and the protection strategy for coastal heritage site from the rising sea-level problem (Attack for Defense) was proposed.

3.2 Data Collection and Descriptive Analysis

3.2.1 Site Problems and Conservation Needs (Macrolevel)

A survey of 209 respondents about the current situation of the Arwad's cultural heritage and the effects of climate change on the Tartous coast were done to locally define the climatic problem. The sample elected the local people who live in Arwad Island for twenty to forty years. Most respondents are housewives females and a few males who work in selling and maritime work. The primary data could be summarized as follows:

① **Climatic Problems and Storms Measures**

Historically, the rundown Phoenician defensive wall, which was built to protect the western seafront of the island from the enemy, high waves, and storms in winter, and the submerged underwater antiquities are two pieces of evidence of the impact of storms and high waves. No great previous precautionary measures against storms and high waves in winter, only the news shortly before about the weather for the fishermen and drivers of the boats to take precautions in order to strengthen the ligament of the ships, monitoring the boats, alerting fishermen, stop the sailboats and close the ports. In case of high wind speed between 30 until 40 km, the ships prevent sail. The sources of weather information are arranged as firstly from local people or family recommendations and experience, then from government weather and national meteorologists, Internet and social networking Web sites, and finally from the international sources and advertisements. A total 80% of the respondents noticed rising sea level now more than before. As well, most of the locals extremely and very concerned about the island's current situation. A total 72% of locals were not satisfied with the government works and strategy for protection from sea-level rising (SLR), and 18% of them were neutral and 10% were satisfied, as shown in Fig. 3.12a–c.

② **Local People Adaptation**

Arwad local people are very simple people, and most of them work in the sea, and most of them have houses or relatives in Tartous, Latakia, and Lebanese coast. They have annual moving from the island to the coasts in bad winter weather. The people who stay in the island store the vegetables, bread, and basic materials in enough quantities before the storm. The citadel area has the highest level on the island above the sea, and it was always the shelters for the local people which they leave in the western seafront of the island from the suddenly high waves and wind in winter. Most locals leave the island for a temporary period in the bad winter climate because of rising sea level and storms, and just 10% of locals do not leave the island because they are from old people category as shown in Fig. 3.12d. Most locals leave for a temporary period in a safe place on the island itself as the citadel neighborhood, their own house in Tartous city. The others go to a relative's house in Tartous city, abroad, or an own house in other city and a slight percent in a relative's house in

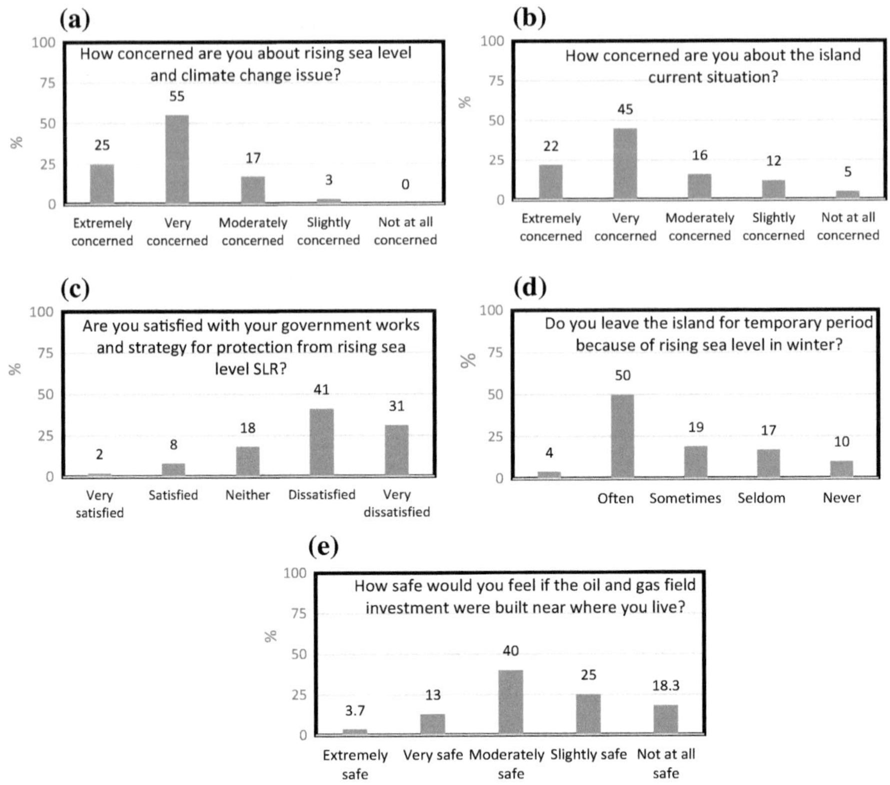

Fig. 3.12 Site survey and the respondents concern about: **a** the sea-level rise and climate change issue, **b** the island current situation, **c** the government works and strategies for SLR, **d** respondents adjust in winter, **e** the oil and gas investment

another city. The passive strategies related to the building's geographical location and design as traditional architecture have been removed in the operating and reuse process, especially with the assistance of the mechanical heating and cooling systems to create an artificial environment to achieve the occupants' thermal comfort. The respondents expressed their extreme willing to change their lifestyle to reduce the damage that the environment causes to the island and their buildings and to share in the process of protecting the island and its building island starting from their houses.

③ Heritage Assets and Site Investment

The most important heritage assets the locals see should be protected and organized by the priority: the historic and residential buildings, the defensive heritage (castle, tower, walls,…, etc.), the ruins and buildings from different historic periods (ancient and prehistoric, Roman, medieval, …, etc.), the religious assets, the civil infrastructures (routes, corniche…), the industrial heritage (boats,…, etc.), the cultural landscapes, and the customs and traditional arts and food. They also considered the negative effects of climate change on the architectural vocabulary of the island, especially the degradation of the building facade of the seafront, the deterring residential use and environmental degradation, and the demolition of Phoenician defense wall and sinking the Phoenician monuments of the island because of high waves. Besides, the negative effects of climate change on the social life of the island on the basis of the locals' personal preferences were the deterioration of the economic and social situation of the families in Arwad has due to lack of resources, the difficulty of securing daily necessities of the consumables and foodstuffs such as vegetables and fruits, the disruption of students from the school in winter because of the difficulty of transferring non-local staff.

The locals determined the needs of the cultural heritage sites to be a destination for tourists and to have a positive effect and prioritized them as following: Create new projects and investments, attract tourists and visitors, create jobs, improve the market for local products, preserve the urban vocabulary and the history, nurture identity of a place, reduce the need for new buildings by promoting the existing buildings, and make it an attractive environment for residents. Oil and gas would be energy sources in the next ten years because of the wind, solar, and nuclear energy are still very weak in Syria. Most locals use heating and cooling systems in their houses during extreme temperatures, and they found that the government should provide more money to support alternative energy and the investment of the oil and gas would be not safe for their life in the island and the safety measures might take place to achieve that as shown in Fig. 3.12e. The Arwadian people would like to see tourist and investment projects on Arwad Island. The authority must take actions to build better planning and strategies started by identifying the priorities and sources then secure the finance and create the suitable investment projects on the basis of locals' opinions.

④ **Conservations Needs**

Despite the 10th five-year plan (2006–2010) and the 11th five-year plan (2010–2015), the Syrian State Planning Commission recommends the need to activate the decentralized approach based on planning from the bottom to guide planning from the top. The proper development and investment projects in the line with resources and heritage preservation needs should be the key priorities of the governments' actions. Based on the questions about the priorities, government role, and decision-makers' opinions, the conservation needs could be summarized as following:

(1) The climatic and demographic challenges for the coastal area could be solved by strategic plans and different solution strategies directed to the specific coastal characteristics and problems; this may encourage a new development for urban growth which contains expected human activities in the future and protect existing coastal areas from SLR problem in accordance with environmental rules.
(2) The need for management and treatment of the historic buildings and dwellings, the cultural built heritage including Phoenician caves, and the underwater archeology and the other type of heritage assets. The need for setting a proper conservation policy and preservation guide of the island and its historic buildings.

(3) Private and public sectors in participation with the local authorities and locals are the most relevant actors in the sustainable preservation and use of heritage assets, then the role of the universities, tourist agencies, managers of heritage assets, restoration professionals, and cultural workers comes later. Regional planning and strategies can foster the local development and determine the economic value of cultural heritage in Arwad Island that are very useful in the future strategies.

(4) Raising awareness and understanding of our heritage and the contribution it can make to everyone's life strengthen the partnerships and resources that support heritage, ensure that heritage plays a full role in strengthening our communities, and ensure that our heritage is conserved for the benefit of future generations are the key priority areas in Arwad Island.

(5) The government decision-makers and planners work individually in a direction and the architect preserve in the opposite direction because of the lack of database, framework, and hierarchical connections of the level of preservation policy and processes without having an in-depth level of knowledge about the local context of the place, local conditions of the environment, materials, and climate, with the absence of preservation vision in the era of danger and international issues.

(6) The basic preservation process in Syria lacks to the research background especially in the environmental field, and local climate impacts on the sustainability of restoration materials and enduring through the time. No significant concern or effective projects put to deal with climate change and rising sea-level problems. Additionally, finding data, and financial resources and professional experts in this field for the decision-makers and projects are not easy, and no mandatory and serious decisions made by the government to support this field.

(7) Setting conservation policy, management strategy, and procedures and the implementation should be in a participation between the involved parties, beneficiaries, and the legislative power such as building owners and managers, archeologists, architects, engineers, surveyors, conservators, planners, conservation officers, contractors, and the local authority.

(8) The monitoring and managing heritage should follow steps and methods started with the comprehensive documenting, diagnosis, and problems identification, then the importance of using simulation programs as assessment tools hand by hand with the exceptional leadership and enhancement visions and management.

In summary, the stone structure of the Syrian built heritage is considered the essential features of the steadfastness of the ancient Syrian civilizations to the present day. The historic Arwad Island and Tartous old city, historical Amrit site and its sectors, and Tartous city seafront are significant heritage sites suffered from climatic problems and need an environmental monitoring and risk management. Based on the data collection, climate history, and site survey and interviews for the local people and employees of the coast guard and Tartous navy, the case study sites face three kinds of climate problems presented by the annual increase of the temperature and humidity, sea-level rising (SLR), and storms and bad weather in winter. A formula of

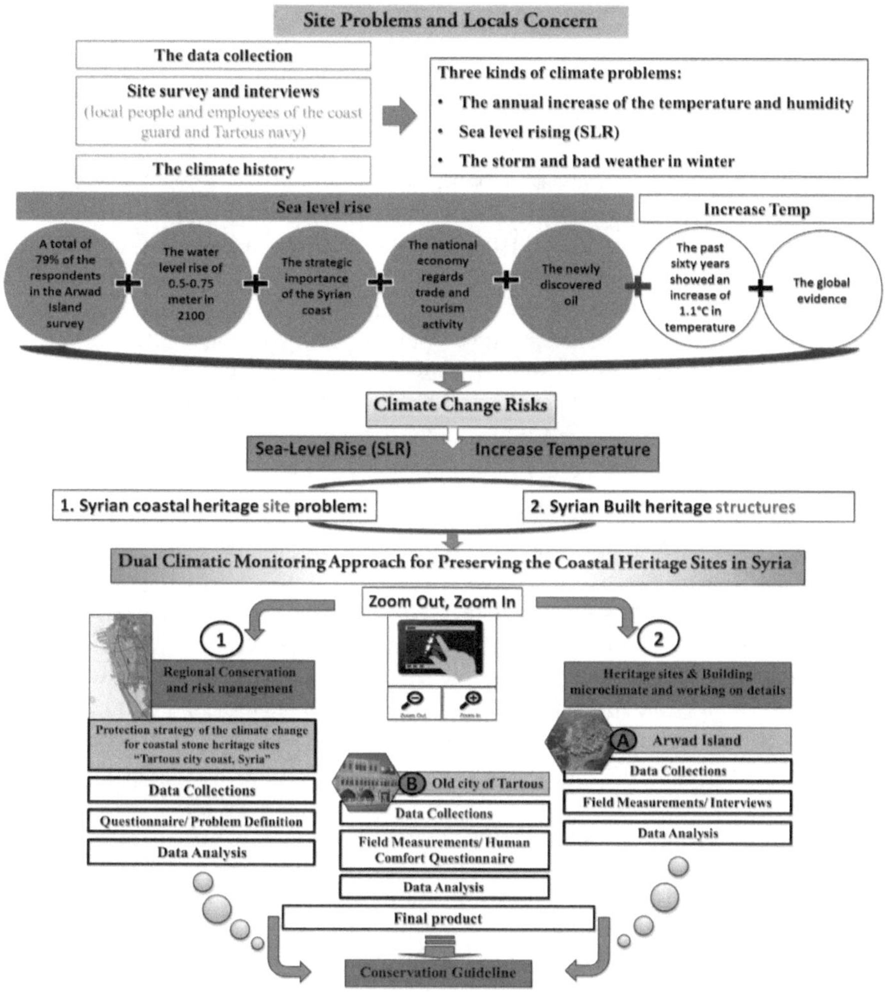

Fig. 3.13 Site problems and preserving vision

the solution has to be sought, starting from the current situation and the growth of the problem, considering that the level of risk is not specified as shown in Fig. 3.13. This book focuses in this chapter on the coastal strategic planning, then later in Chaps. 4 and 5 on documenting the historic structures and providing a subjective and objective dataset of the current situation of the built thermal environment of the two inhabited threatened sites structures in Arwad Island and the old city of Tartous and the users' behavior. Where research in this direction is not found, that makes any research in this direction a big contribution.

3.2.2 Tartous Coast and Syrian Continental Shelf's Possibilities

① The possibilities of the Syrian continental Shelf

Based on utilizing the survey of Commander Mansell and French Government Maps of 1952 for the topography of Mediterranean Sea for Syria and Lebanon from Latakia to Sour, the map of Arwad Island port of 1960, the map of the ports in Syria-D-Tartus of 1999, and the data collection from Maritime Office of the port and local navigators of Tartous , a clear image of the continental shelf opposite of Syrian coast and marine depths was obtained starting from the port of Tartous in the north to the Syrian–Lebanese border in the south (Mansell 27 Aug. 1954, Hydrographic office of Admiralty 26 Aug. 1960, Admiralty 25 Feb. 1999). From the analysis of the possibilities of the continental shelf opposite the Syrian coast and detailed analysis of the marine depths corresponding to the slice studied, we found:

(1) Syrian coast characterized by good space within the Continental Shelf, Fig. 3.14. There is an underwater mountain range running parallel the coastline of the eastern basin of the Mediterranean.
(2) The good extension of the continental shelf in front of the Syrian coast gives an idea about the gradient acceptable marine depths.
(3) The existence of a rocky floor paralleled the Tartous shoreline, starts from Arwad Island, the only inhabited island among a series of small islands, and continues to other small islands and spots sequential toward the south, as well as another set of submerged rock masses under water between this range and the shoreline as shown in Fig. 3.15.
(4) Arwad Island and the nature of the geological rock for many parts of the continental shelf give us a clear vision of the focal points of the supposed defensive line, and the possibility of creating a collection of artificial defensive islands built on the rocky apparent bases in the marine survey schemes with proportionate heights for the problem of SLR. Then grows continuously according to time strategy plans to merge in the second stage in one defensive line and follow the growth direction of shoreline in the third stage based on specific axes in advance, taking into account the sea current movements in the study area. In each stage of growth, leave the set for the continuation of the ports, during periods of growth before preparing full-lock that comply with the growth of the problem of SLR, to the extent that it constitute the maximum risk to human communities on the shoreline, as the problem of rising water is problem non-momentary will gradually increase over time.

② The Current Situation of Tartous Coast

Arwad Island currently suffers from the storms and high waves' problem that cause temporary emigration of the local people every winter from the western waterfront house to the center of the island or the city of Tartous, without any measure or protection project since the Phoenician great defense wall that still exists even today

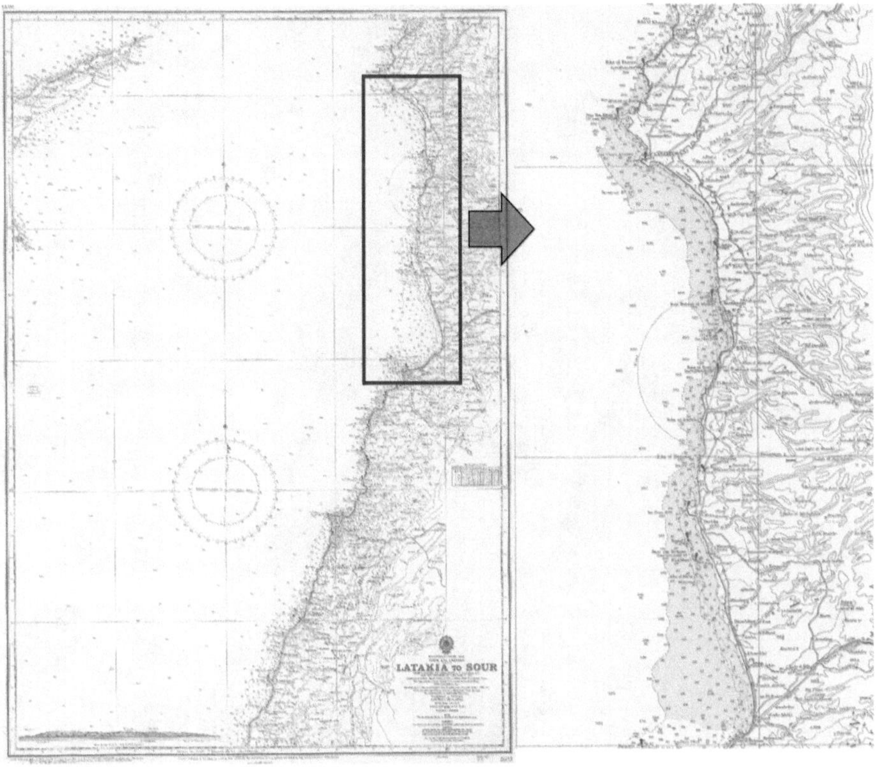

Fig. 3.14 Mediterranean Sea Syria and Lebanon topography. *Source* Mansell 27 Aug. (1954), Admiralty 25 Feb. (1999), Rahmoun et al. (2016)

as remnants. The submerged ruins underwater that mentioned by the archeologist Honor Frost confirms the historical SLR problem. Tartous City stayed a long time as an important agricultural city; the waterfront contains the old city and the city center which also remained without development projects. Due to the importance commercial and tourist role, Tartous municipality started developing the real estates and tourism of the shoreline since the beginning of this century by two reinforcement operations and urban expansion area projects. The first project is the promenade seashore of Tartous municipality for public entertainment that has completed since 2006 with a group of public facilities, squares and areas of the tourism investment with arms extended into the sea as a starting point toward the defense plan. In 2008, the second project is the 7-star tourism investment project behind the Tartous port began working as shown in Fig. 3.16. All engineering and construction works of seashore projects in the parts implemented or being implemented are designed to protect the residential area of the city of Tartous from high marine waves, which is located at the same time under the risk of SLR. Strong winds that have occurred recently in 2014 and 2015 proved that the seafront promenade development project is not designed

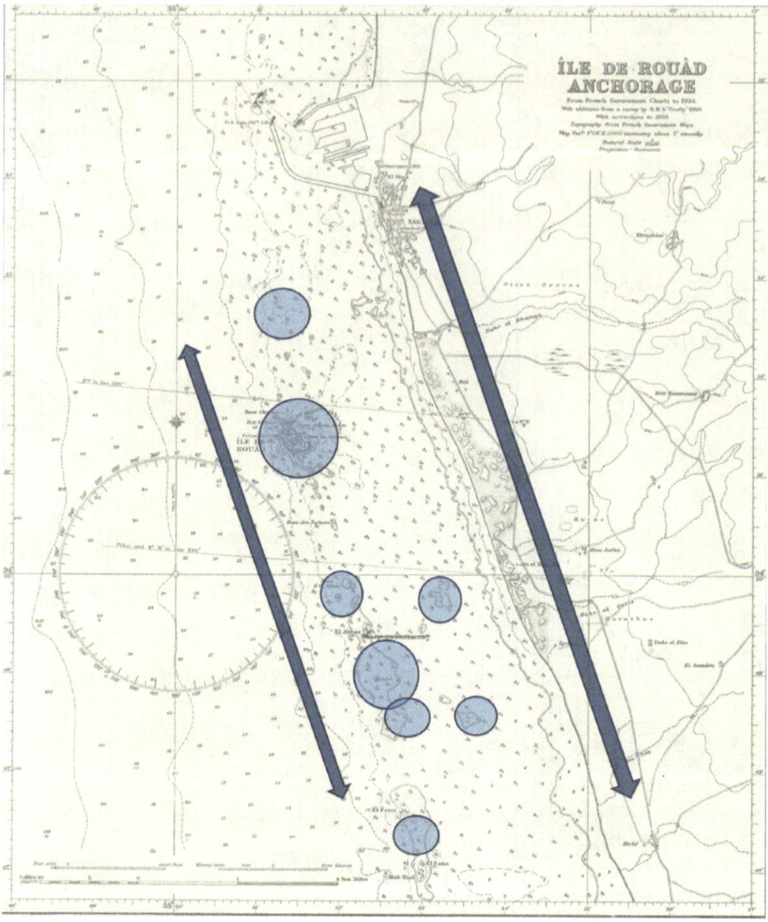

Fig. 3.15 Analysis of the possibilities of the continental shelf. *Source* Hydrographic office of Admiralty 26 Aug. (1960), Rahmoun et al. (2016)

to be the defense line of the waterfront sector, and the property development of the shoreline, which proposed by the municipality to be the new waterfront of Tartous city, did not notice the SLR problem. The same situation for Amrit site and its futuristic tourism development projects in the southern sectors of the city of Tartous. Studying the emergence and development of the city of Tartous guides the proposal optimized to exploit the defensive islands sustainably. The public project of Tartous municipality could be considered as the arm that will carry the proposed final defenses in the first step, in addition to that it gives a clear idea about the important part of the nature of human activity (tourism activity) for supplementing tourism role of the city, when studying sustainable planning process blue communities (defensive islands able to grow).

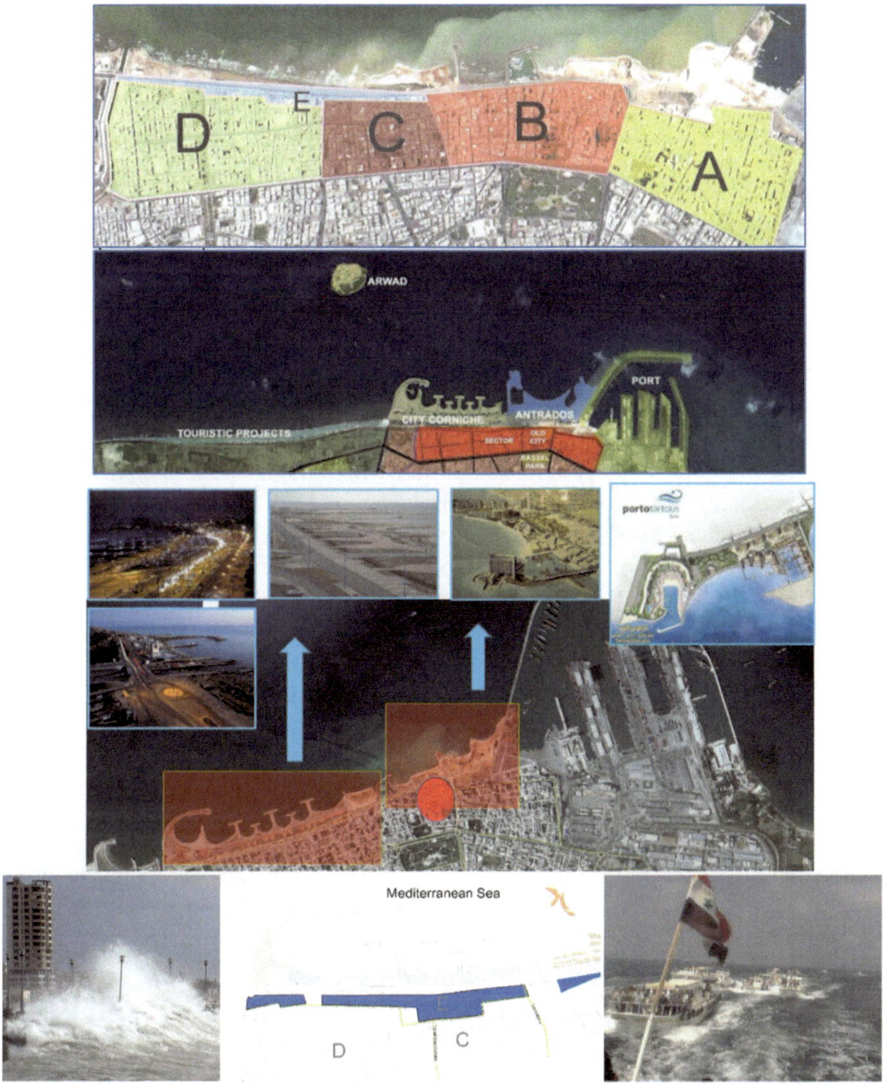

Fig. 3.16 Tourism development projects, on the coastline within the study case area. *Source* Tartous Municipality, Urban Planning section

3.2.3 Syrian Territorial Water's Possibilities and the Investment Plan

① **The Possibilities, Sources, and the Investment Plan**

(1) In 2007, ANSYS, Inc Co., conducted a survey of Geodesy Syrian coast, stretching from the Lebanese border north to the Turkish border, including various marine depths, according to the findings of the competent Norwegian company also analyzed the information and data, and confirmed that there are 14 oil fields in Syria.
(2) Modern American studies recently confirmed the discovery of gas and oil treasure in a huge basin of the Mediterranean estimated reserves to 122 trillion cubic feet of natural gas and 107 billion barrels of oil (Schenk et al. 2010).
(3) In 2011, the Ministry of Petroleum and Mineral Resources launched an international tender for the work of prospecting, exploration, and exploitation of oil and gas. The announcement includes a presentation of three marine areas with an approximate area of about 3000 km^2 for one block (Ministry of Petroleum and Mineral Resources 2011; CGGVeritas 2011), as shown in Fig. 3.17.
(4) The four largest oil fields are concentrated off the Syrian coast and down to the city of Banias and are expected to be equivalent to discovered quantities daily production up to 1.6 million barrels per day (Shuaibi 2013).

All the above gives a clear image for the future new human activity on the Syrian coast and provides an idea of sustainable planning in the future to the Syrian coast in general and the importance of finding a suitable areas for the presence of investing companies, proposed defense islands to protect the coastline of the headquarters appropriate for this new future activity allowing trade and economic investment opportunities for this defensive islands, as well as tourism investment and reducing their impact on the heritage sites.

② **Facts and Scenario Background**

Urantia Book gives specific details about the sinking of a peninsula off the eastern coast of the Mediterranean some 33,000 years ago (Urantia Foundation 1955), Fig. 3.18. At that time, the theory of continental slipping and the sinking of a section of land in the water and scanning the marine depths was not yet known, as does the navy sonar images. However, the modern technologies for marine survey currently reveal a topography with small depth under water off the Syrian coast (Halbert Katzen 2007), as shown in Fig. 3.19. On the other hand, according to the latest scientific theories at the moment and to the archeological study of Phoenician Arwad Island by a joint Syrian–Japanese mission in 1985, Arwad is not entirely a natural island, but it is considered an artificial island built purposely in the sea to increase the habitable land areas on the expense of the sea, using marine construction methods as is the case in the construction of the artificial islands and subsequent ports (Hijazi 1991).

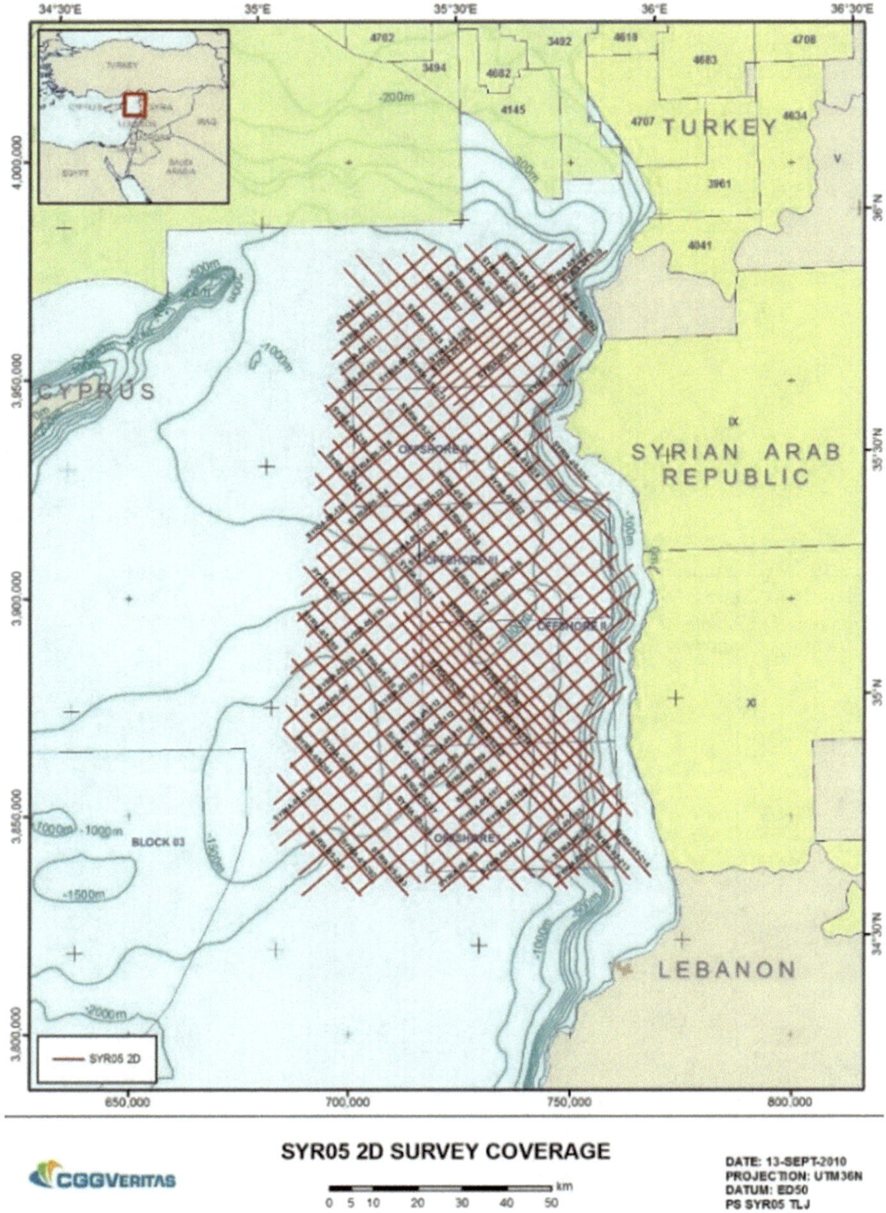

Fig. 3.17 CGGVeritas 2D survey coverage offshore Syria. *Source* CGGVeritas (2011)

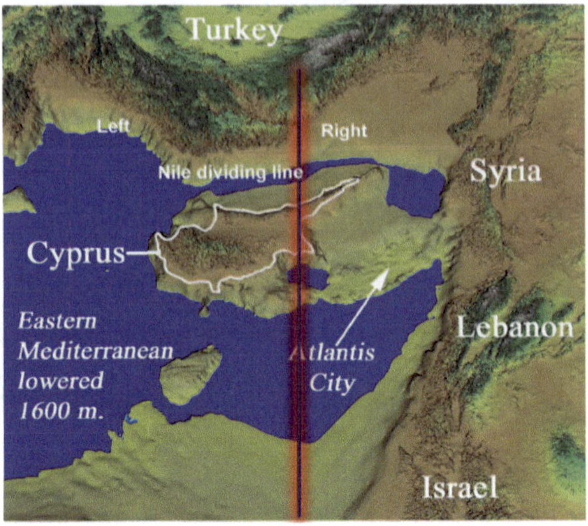

Fig. 3.18 Pangaea to plate tectonics report. *Source* Halbert Katzen (2007)

Fig. 3.19 Mediterranean Seafloor Map, National Geographic 1982. *Source* Halbert Katzen (2007)

Previous documents, with a group of tourism projects put forward by the Ministry of Tourism in 2006, based on marine backfill, and artificial islands to support tourism investment and after comparing data for Syrian coast with global scenarios in the face of SLR danger. Thus, the idea was launched mainly from the supposed possibility to restore parts of the continental shelf that submerged under water off the coastal cities on the Syrian shores back to life again, with a new structure suitable for the marine environment; constituting at the same time the real defense of the coastal population in the face of SLR problem. The three steps of analysis and continental shelf, topographic survey schemes and marine depths, in parallel with the analysis of the nature and the needs of current communities, specify the nature of the advantage of defenses (artificial islands viable to connect in the future as a series of islands with continued growth), and places most convenient for its establishment. In order to develop sustainable development plans, and suggest the nature of the proposed human activity in the future plans, between the shoreline and the defensive line, as well as the appropriate directions for future growth.

3.3 The Proposed Scenario (Attack for Defense)

3.3.1 Discussion of the Global Scenarios Facing SLR Risk

The review of global scenarios for the proposed solutions in the face of danger SLR, during the different periods of time (2030–2100) in various places around the world and in different environmental impressions depending upon the speed of the increase expected in sea-level rise and topography situation, led to many practical strategies and scenarios which give different solutions from no action to construction the Levees or retreat, to protect the historical and current human inheritance, and secure safe communities for life in the future (McNamee et al. 2014; Tol et al. 2006), as shown in Fig. 3.20. The global scenarios with three different visions could be summarized as:

- **Retreat**: Retreat from threatened locations and pull out toward the areas of the desired altitude to face the risk of flooding, the supporters of this scenario find the idea of sustainability through new habitats resulting from the retreat. This scenario requires coastal areas with low population density that will not be available with the natural increase in population or environmental migration, as well as the loss of authenticity of the heritage buildings by moving them from their locations.
- **Defend**: To protect the residential areas from entering the water by the future growth of cities toward the water bodies in order to face the SLR problem. Building defenses costs are not acceptable without additional advantages in providing alternative energy such as wind power and wave power could be invested in developing the system of the coastal area. However, these defenses will remain limited to the future growth of cities toward the water bodies.

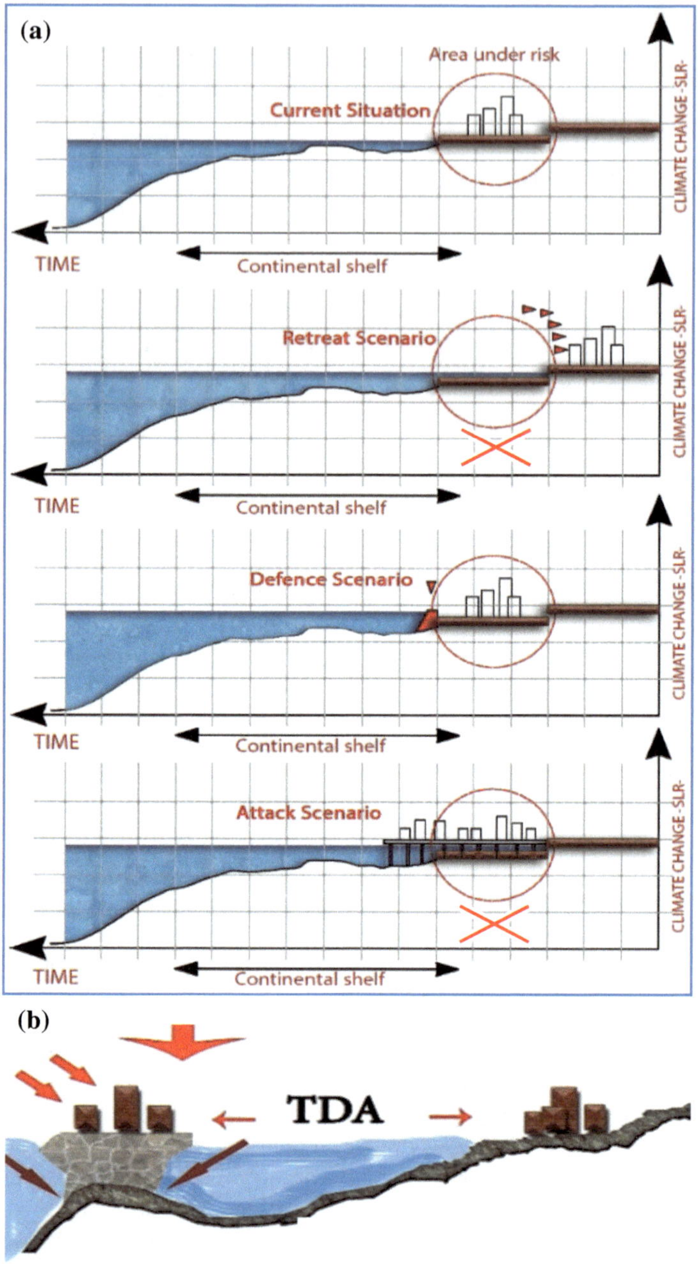

Fig. 3.20 **a** Global scenarios in the coastal area against SLR, **b** the consolidation scenarios in a new one

- **Attack**: Real steady progress toward the water by using new technologies that have been utilized in the twenty-first century as the floating structures, construction of homes with moving solutions appropriate to the problem of high water levels, and even move toward the floating cities, and initialize new residential areas to receive water according to good plans. This requires enormous economic costs with no real guarantees for the protection of existing residential areas and vital facilities (Hardiman et al. 2010).

3.3.2 *Protection Strategy of the Climate Change for the Coastal Heritage Sites in Syria*

Based on the discussion of the global scenarios in the coastal area facing SLR risk (Retreat, Defend, Attack), and referring the theoretical background in the chapter two, with the linkage to the heritage value and authenticity, the proper heritage conservation scenario could be determined. The stone material is considered the most important features of the ancient Syrian civilizations in most heritage sites, and helped to the steadfastness of these civilizations, despite all the natural and human conditions to the present day, and achieved the spiritual relation between time and space where they built on. Therefore, the stone structure and construction of the built heritage are sensitive for applying the previous protection theories and solutions, where it is impossible to transfer this stone historical coastal sites which threatened by the SLR problem to some other areas, without losing their authenticity and cultural value, which is linked to the sea and the transition between the world beaches. The proposed scenario called (Attack for Defense). The new particular scenario to protect the threatened coastal cities from the SLR danger on the Syrian coast, based on consolidation scenarios attack and defense in a new idea, is the attack and stability for defense in the future, which is based on sustainable planning of the coastal region. Because the risk of SLR problem starts from the erosion of beaches gradually stage, until reaching the immersion stage, as shown in Fig. 3.21.

Develop the plans and defensive strategies compatible with the nature of the coasts that are threatened in each region of the world, which is the key solution to address SLR and the increasing of the demographic problems in the coast, taking into account the environmental, demographic, and ecological impacts in the area of the application scenario. The targeted area in this scenario is characterized by a natural potential of the continental shelf and allows the application of an advanced defensive scenario for the threatened historic area by building a series of scattered artificial islands in front of the coast, which are available to grow gradually in line with time and the exacerbating of the problem of rising sea levels to the extent that it constitutes the maximum risk to human communities on the shoreline. This proposed scenario will lead to the emergence of a group of designed islands, taking into account the problem RLS and the possibility of expansion and integration according to directions imposed by the study of the continental shelf, through a specific planning stages which constitute the

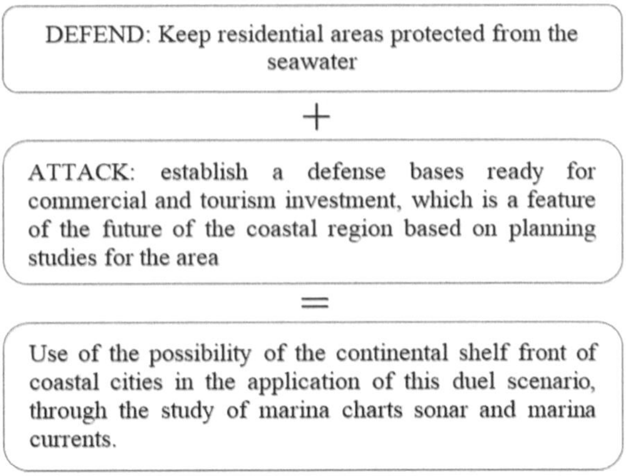

Fig. 3.21 Idea of the proposed scenario (Attack for Defense)

islands in the future shield that will protect coastal cities in the previously a specific points, so as to give a solution, taking into account the expansion of these cities. It inevitably ECO-LANDS that will accommodate human activity in the future, selected for each coastal city according to foundations of regional planning, but within the rules of blue communities compatible as much as possible with necessary marine environmental requirements, such as the use of alternative energy and new applications in waste disposal, with modern environmental science ways (Salahuddin 2009; Mangor 2004), as shown in Fig. 3.22.

Then merging the islands in the second stage in one defensive line and following the growth direction of the shoreline in the third stage based on specific axes in advance, taking into account the sea current movement in the study area, to reach at the end the implementation of the scenario to what looks like an arched shield in front of the protected area. At the same time, these islands as they grow naturally could receive the increased human activities in the coastal region, thus as well as the defensive role they could have formed a new urban spaces, so the required environmental requirements should be put before, during, put up for investment within the concept of ECO-LANDS and blue communities. On the other hand, the gradual growth process of the defensive arc process toward the opposite coasts must be synchronized with the documentation survey of the submerged human heritage in the study area and to launch the preservation, maintenance, and rehabilitation processes during the implementation of the defensive scenario.

3.4 Summary

A series of data collection was applied in this chapter. The lack of environmental research and the absence of sustainable regulations and monitoring systems in Syria were presented. This chapter survey provided grounded information on the current

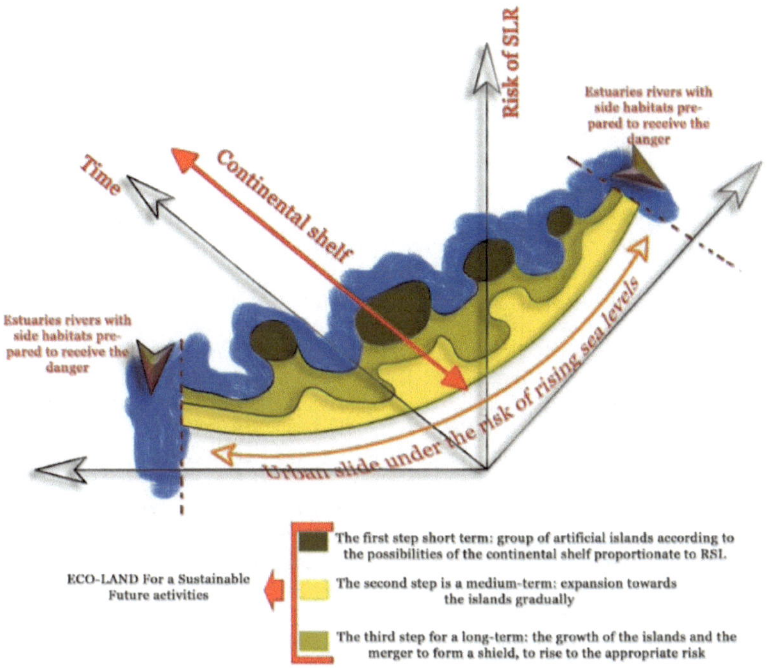

Fig. 3.22 Proposed scenario

situation of the Arwad's cultural heritage and the effects of climate change on the Tartous coast in Syria based on the participation of the island locals, the people who concern on, and the stakeholders. Therefore, the annual increase of the temperature and sea level on the Syrian coast, and the threatened the coastal heritage sites in Tartous city were theoretically and locally identified. The procedures and steps for the development of the plans were determined. The development of appropriate sustainable defensive strategies for the affected coastal areas due to climate change was demonstrated. Analysis of the information collected for the possibilities of the continental shelf of Syrian coasts to create defensive islands able to grow in defense system according to the scenario (the Attack for Defense) based on the ECO-LANDS requirements. Protection of minatory coastal areas of the main problem of climate change (SLR) was proposed, taking into account the environmental, demographic and ecological impacts in the area of the application scenario, as a strategic shift from a reactive to a proactive preventive approach.

References

Abdulkarim, M. (2013). *The archaeological heritage in Syria during the crisis*. Damascus: Ministry of Culture Directorate General of Antiquities & Museums.

ADEME. (2010). *Thermal Standard for buildings in Lebanon (TSBL). For new residential and non-residential buildings*, s.l.: s.n.

Admiralty, B. (1999, February 25). *Mediterranean sea, Ports in Syria, D Tartus Panel [Map].1:25,000. BA1579. INT.3669*. BA1999 ed. Taunton, United Kingdom: British Admiralty under the superintendence of Rear Admiral J P Clarke, CB, LVO, MBE, Hydrographer of the Navy.

Adorni, E., & Venturelli, G. (2010). Mortars and stones of the Damascus Citadel (Syria). *International Journal of Architectural Heritage: Conservation, Analysis, and Restoration, 4*(4), 337–350.

Aga Khan Award for Architecture. (1991). *Sweida elementary school in Sweida "Shagrawieh", Award No 1238.SYR*, Sweida. Syria: Aga Khan Award for Architecture.

Al Dakar, A. (1999). Treatments by radiation control of roofs at the Syrian coast (In Arabic). *Journal for Research and Scientific Studies of Tishreen University, 21*(8)

Al Dakar, A. (2000a). Climate and the urban ecology at the Syrian coast (In Arabic). *Journal of Tishreen University, Lattakia*.

Al Dakar, A. (2000b). Using of gyps in the thermal treatment of the roofs by radiation (In Arabic). *Journal for Research and Scientific Studies of Tishreen University*.

Al Dakar, A. (2005). Traditional architecture and cross ventilation in Lattakia (In Arabic). In *Urban and Architectural Heritage Symposium in the Province of Lattakia, Lattakia*.

ArchDaily. (2017). *6 endangered world heritage sites as seen from space*. (Online) Available at: http://www.archdaily.com/873265/6-endangered-world-heritage-sites-as-seen-from-space. Accessed July 20, 2017.

Attallah, S. O., Senouci, A., Kandil, A., & Al-Derham, H. (2013). Utilization of life-cycle analysis to evaluate sustainability rating systems for construction projects with a case study on Qatar Sustainability Assessment System (QSAS). *Smart and Sustainable Built Environment, 2*(3), 272–287.

Bashour, N. (2011). *Documentation of the old city of Tartous*. Damascus: Tartous Governorate in cooperation with the Municipal Management Modernization Program (MAM), funded by the European Union

Cassar, M., & Pender, R. (2005). *Climate change and the historic environment*. London: University College London with the English Heritage and United Kingdom Climate Impacts Programme.

CGGVeritas. (2011). *CGGVeritas offers geophysical and geological datasets to support the Syrian international offshore 2011 bid round*. (Online) Available at: https://www.cgg.com/en/Media-and-Events/Media-Releases/2011/03/cggveritas-offers-geophysical-geological-datasets-to-support-the-syrian-international-offshore-2011-bid-round. Accessed March 18, 2013.

Church, J., et al. (2013). *The physical science basis. Contribution of working group I to the 5th assessment report of the intergovernmental panel on climate change*. United Kingdom and New York, USA: PM Cambridge University Press.

Climate Directorate in Damascus, from 1950 to 2012. *Annual climatological data*. Damascus: Climate Directorate, Meteorological Department, Ministry of Defence, Syria

DGAM. (2017). *Ministry of culture: Directorate general of antiquities & museums (DGAM)*. (Online) Available at: http://dgam.gov.sy/index.php?d=314&id=2280. Accessed July 20, 2017.

Faour, G., Fayad, A., & Mhawej, M. (2013). GIS-based approach to the assessment of coastal vulnerability to sea level rise: Case study on the eastern mediterranean. *Journal of Surveying and Mapping Engineering, 1*(3), 41–48.

General Directorate of Meteorology in Syria. (1977). *Climate atlas of Syria* (150 pages). Damascus: General Directorate of Meteorology in Syria, Climate Directorate, Military survey management.

GISS Surface Temperature Analysis. (2014). *Lattakia station*. (Online) Available at: http://data.giss.nasa.gov/cgi-bin/gistemp/show_station.cgi?id=647400220000&dt=1&ds=14.

Halbert Katzen, J. (2007). *Pangaea to plate tectonics*. (Online) Available at: http://www.ubthenews.com/topics/pangaea.htm#report. Accessed March 18, 2013.

Hardiman, N., et al. (2010). *Facing up to rising sea levels: Retreat, defend, attack?*. London: Building Futures and Institution of Civil Engineers.
Hassan, M. (2006). *Differentiation of bioclimatic regions in Syria and its effect on architectural designs* (Diploma of high studies's thesis, Faculty of architectural engineering, Tishreen University, Lattakia, Syria)
Hassan, M., & Rahmoun, T. (2016). *Protection strategy of the climate change for coastal heritage sites in Syria, China* (pp. 474–483). Nanjing: Digital Media Press of Nanjing University.
Hassan, M., Xie, H., & Rahmoun, T. (2017). *The current built environment of the medieval forts on Arwad Island, Syria*. Cádiz, Spain, s.n.
Haydar, F., & Pediaditi, K. (2010). Evaluation of the environmental impact assessment system in Syria. *Environmental Impact Assessment Review, 30,* 363–370.
Hijazi, S. H. (1991). *Arwad Island and Phoenician roots*. Damascus: Dar Al Amani.
Hydrographic Office of Admiralty. (1960, August 26). *Île de Rouâd Anchorage. From French Government charts to 1924. With additions from a survey 1860. With corrections to 1950 [Map].1:37,500. 1036. A.1-77.* Engraved 1960, under the supervision of Rear Admiral E.C. Irving. O.B.E. Hydrographer ed. London: Admiralty.
Kanter, R. M. (2011). Zoom in, zoom out. *Harvard Business Review, 89*(3), 112–116, 131.
Komurlu, R., Arditi, D., & Gurgun, A. P. (2014). Applicability of LEED's energy and atmosphere category in three developing countries. *Energy Build, 84,* 690–697.
Komurlu, R., Gurgun, A. P., & Arditi, D. (2015). Evaluation of LEED requirements for site properties in developing country-specific certification. *Procedia Engineering, 118,* 1169–1176. Chicago, USA.
Mangor, K. (2004). *Shoreline management guidelines* (3rd ed.). Horsholm, Denmark: DHI Water & Environment.
Mansell, A. (1954, August 27). *Mediterranean sea, Syria and Lebanon, Latakia to Sour (Map).1:300,000. 2633. A.11-73.* Enamel 1954 ed. London: Admiralty Sailing Directions, under the supervision of Vice-Admiraly Sir Archibald Day, K.B.E., C.B., D.S.O., Hydrographer.
McNamee, K., et al. (2014). Scenario planning for building coastal resilience in the face of sea level rise: The case of Jacobs Avenue, Eureka, CA. *Humboldt Journal of Social Relations, 36,* 145–173.
Ministry of Petroleum and Mineral Resources. (2011). *International tender for the work of prospecting, exploration and exploitation of oil and gas resources in some areas of territorial sea and exclusive economic waters of the Syrian Arab Republic in the Mediterranean*. Damascus: Ministry of Petroleum and Mineral Resources and GPC.
Nakicenovic, N., et al. (2000). *Special report on emissions scenarios a special report of working group III of the intergovernmental panel on climate change*. Cambridge, New York: Cambridge University Press.
NASA Goddard Institute for Space Studies. (2014). *GISS surface temperature analysis, Latakia Station.* (Online) Available at: http://data.giss.nasa.gov/cgi-bin/gistemp/show_station.cgi?id=647400220000&dt=1&ds=14. Accessed December 11, 2015.
Nezam, H. (2004). *Developed ways in restoration the stone used in the historic buildings of the Syrian Coast* (Master's thesis). Lattakia: Tishreen university.
Olgyay, V. (1953). *Bioclimatic approach to architecture.* s.l., BRAB Conference Report, (Vol. 5), National Research Council, Washington, DC.
Rahmoun, T., Hassan, M., & Alhasan, W. (2016). Protection strategy for the coastal areas of climate change "Syrian Coast, Tartous City". *Research Journal of Applied Sciences, Engineering and Technology, 12*(3), 264–271. ISSN: 2040-7459.
SASMO. (2015). *Engineering standards.* (Online) Available at: http://www.sasmo.org.sy/ar/search_standards. Accessed January 15, 2016.
Salahuddin, J. S. (2009). *Coastal management Volume 3.* Kuala Lumpur: Government of Malaysia Department of Irrigation and Drainage (DID Manual).
Schenk, C. et al. (2010). *Assessment of undiscovered oil and gas resources of the levant Basin Province, Eastern Mediterranean.* (Online) Available at: http://energy.cr.usgs.gov/.

Shuaibi, I. F. (2013). *Oil and gas conflict globally and regionally* (Interview) (1 4 2013).
Tol, R., et al. (2006). Adaptation to five metres of sea level rise. *Journal of Risk Research, 9*(5), 467–482.
UNDP/GEF and MPWT/DGU. (2005). *Climatic zoning for buildings in Lebanon,* s.l.: s.n.
UNEP/MAP-METAP SMAP III Project. (2009). *Syria's coastal zone and its desired integrated management*. s.l.: Priority Actions Programme. Regional Activity Centre.
Urantia Foundation. (1955). *The Urantia book.* Chicago: Urantia Foundation.

Chapter 4
Microclimate Field Survey in Arwad Island, Syria

This chapter presents a simple step in the Zoom in method. It highlights and investigates the thermal environment of two typical defensive stone heritage buildings, the citadel and tower in Arwad Island, that have been reused as museums. The heat and humidity harm the building materials and the exhibits and cause discomfort and visible damage, presented by the green stains and mold on the walls, especially on the ground floor. The study adopted data collections, analyzed a series of field measurements of the summer hygrothermal parameters, and interviewed local residents to identify the characteristics of the historical buildings and to complete an indoor thermal investigation. This chapter aims to study the current built environment and provides essential datasets of the microclimate factors of the built heritage as a potential risk for conservation, and to explore the appropriate control strategies to achieve climate management during the hot season and within the tight budgets of museums and historic buildings.

4.1 Overview of the Case Study Sites

Arwad Island is located 3 km from Tartous shoreline with 27 ha of the residential area in Syria. The total population of the island is 23,000 people; the current resident population is not more than 7000 people (840 families), while the rest of people generally live and work in Syrian coastal cities and abroad (Land Identification and Registration Committee 2014). Arwad's history begun in the early second millennium BC when it was a strong Phoenician kingdom called Arvad (Dumper and Stanley 2007). Arwad spread its influence on most of the Syrian coast on the Mediterranean Sea as a base for maritime trade and built a Phoenician defensive wall to protect the western seafront of the island from the enemy, high waves, and storms in winter. Arwad Island forms a unique architectural, cultural and urban heritage, due to a combination of geographical and historical aspects, and its strategic position for trading and later for military purpose throughout the ages. The importance of Arwad during the Crusade-Muslim conflicts is testified by the presence of two fortifications,

M. Hassan and H. Xie, *Climate Change and Conservation of Coastal Built Heritage*, https://doi.org/10.1007/978-981-13-8672-5_4

the citadel and the tower. The General Directorate of Antiquities and Museums in Syria recorded it as a national heritage site in 1975, and it has been submitted to the tentative World Heritage list of UNESCO in 1999 (UNESCO 2016). The island now is only a small town dependent mainly on local daily tourism with lack of modern amenities. The cultural heritage buildings of this island need annual treatment and preservation, but less research has been undertaken into Arwad. The conservation of Arwad's heritage needs a treatment to include a wide scope of the site action, strategy and ending with control of the internal built environment and details (Živković and Džikić 2015; Balocco and Calzolari 2008; Rahmoun et al. 2016). To get as close as possible to the sustainable conservation, the first aspect of environmental control will be the building envelope, which is the first line of defense against the climatic parameters that affect the heritage buildings and an integral part of managing the climatic limits which are needed for conservation of the collection and the buildings. The second aspect will cover site concerns requiring actions to reduce energy costs; such as changes to the lighting, fenestration and operating hours. An approach for microclimate risk analysis must be defined to set the limits on the acceptable environment in conjunction with heritage and conservation considerations. Therefore, the study of microclimate for heritage buildings is a novelty in comparison with the current scientific literature. The classification of Arwad heritage structures can generally be divided roughly into five types as following, Fig. 4.1:

① The significant remains of Phoenician massive defense wall facing the sea
② Phoenician submerged underwater antiquities
③ The old rock ruins and caves carved in the rocks in the western facade of the island
④ Arwad archaeological buildings represented by the citadel, tower and the remains of the Ottoman bath
⑤ The traditional residential buildings belong to various periods.

4.1.1 Arwad Medieval Forts' Structure and Design

The citadel and tower of Arwad expertly blended the imported and local ideas, materials, and technologies, that make them an example of indigenous creativity (UNESCO 2016), Fig. 4.2. Both buildings were registered as national heritage buildings in 1959. The citadel location rises 14 m from the sea and it was likely a former Phoenician temple (Nicolle and Hook 2004). In the construction of the fortifications; the builders naturally made use of the walls and foundations of the ruins of another existing structure, then they added new walls. It is unclear if these fortifications were built in the Crusader period or the Islamic period, but the construction style of both periods can be seen in the tower and the citadel. The cross-vault construction and the Crusader inscription of Lusignan family in the medieval period are on the right and left of the main gate of the tower and the citadel, as well the vaults and domes of the Ayyubid and Mamluk architecture style, as shown in Fig. 4.3.

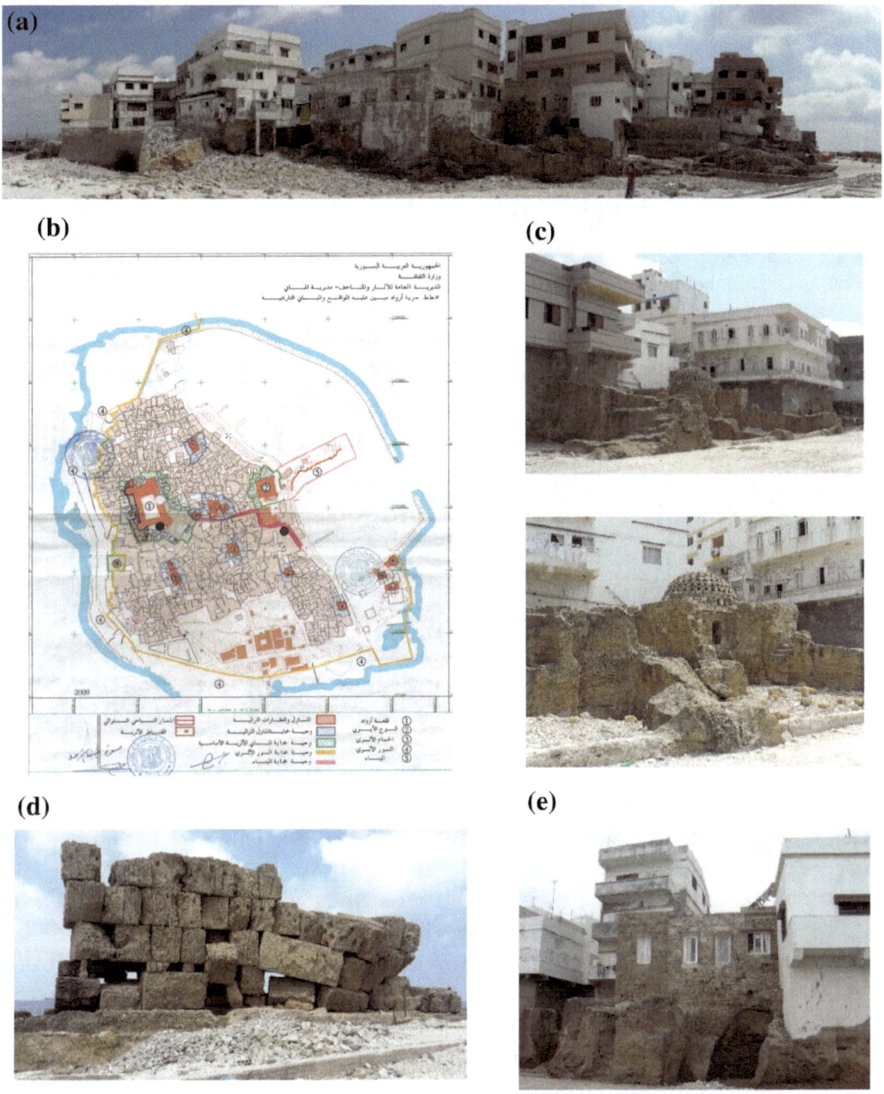

Fig. 4.1 Arwad Island' heritage structures: **a** Arwad Island part of the waterfront, **b** Island plan, **c** Ottoman bath, **d** Phoenician defense wall, **e** rock caves. *Source* Author (**b**): Antiquities and Museums Department in Tartous (2015)

Fig. 4.2 Case study sites: **a** Arwad Island, **b** Arwad citadel, **c** Arwad tower. *Source* Antiquities and Museums Department in Tartous (2015)

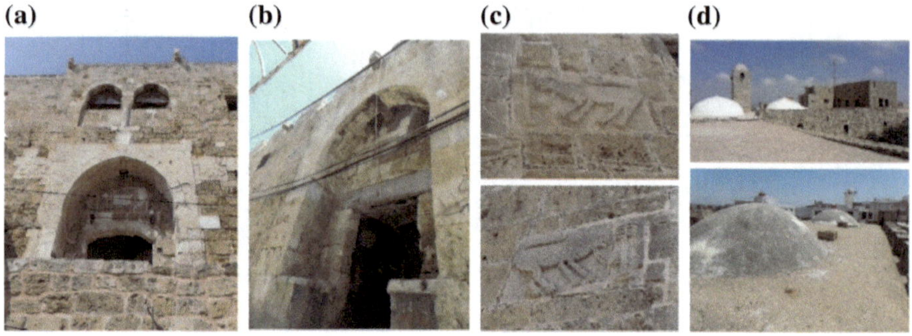

Fig. 4.3 Features of the case study sites: **a** Citadel entrance. **b** Tower entrance. **c** The inscription of Lusignan family. **d** Arabic elements in the citadel (above) and tower (below)

These two buildings were chosen to be measured because they met two important criteria: they are the most significant buildings on the island that are still in use. They were initially constructed for defensive purposes but have had various uses throughout the subsequent centuries and currently they serve as a museum. According to 2003 ASHRAE Handbook, the most traditional buildings that reused as museums in Syria could be classified under Class 4 as heavy masonry structures, typical of high-quality historical houses (ASHRAE 2003). The research study benefited from the use of these buildings for two reasons; they are easier to access than private residence buildings because of the privacy of the island demographic and the locals' distrust of strangers especially with current Syrian war. The second reason is that necessary data and architectural drawings of these two buildings were available in the Directorate General of Antiquities and Museums in Tartous, as was the previous study of rehabilitation of Arwad Island in 2005 (Hassan and Rahmoun 2005).

① Arwad Citadel

The primary function of building the citadel was to serve a defensive purpose, but its function changed throughout time, as displayed in Fig. 4.4. The ground floor of the fortress rises about 2 m above the street with the defensive massive load-bearing sandstone walls. A side staircase leads to the entrance by a main and lonely gate, the entrance leads to the roof by stone stairs on the right side of the gate. The door in

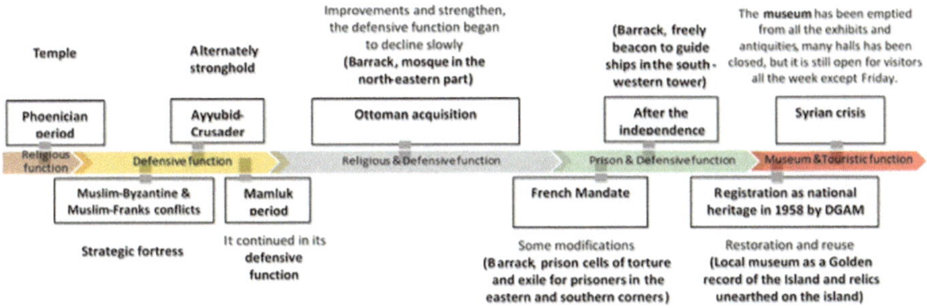

Fig. 4.4 Changing of the citadel functions through the time

front of the gate is open to the main inner courtyard, which is surrounded by several halls with vault construction of sandstone. The halls were used by the inhabitants of the building for different purposes. Many halls are open to the courtyard and to the outside at the same time. Some of them contain fireplaces constructed inside the walls for heating, while some halls have wells dug in the rock to collect water from the roof by gutters; pottery extends within the walls. The main inner courtyard leads by an internal gate to another rectangular yard in the north which connects between the northwest tower and the northeastern tower and works as a defensive part of the citadel. The northern inner courtyard is fortified with walls which rise about 6 m and is equipped with the openings for arrows, stone gears, and gangway of 1 m width for the bowmen. Four towers are located in the citadel corners to increase the immunization; three of them with a circular shape and the fourth is rectangular, and all of them are equipped with the opening of arrows, heated oil, and stone gears. The first floor does not encompass the entire space of the roof, and it has a large room above the entrance of the citadel with two small rooms beside it. It perhaps was the throne room of the governor of the castle in ancient time and has a good view of the whole citadel, island and Tartous coast. The first floor also contains one separate room and two small rooms above the hall of the southwestern tower, Fig. 4.5.

The museum contains glass and pottery collections that were discovered from the sea, the old port and places that belonged to the Island, a set of Venus handcuffs, a panel for stereoscopic Phoenician ship, models of sailing boats which were used until recently, fishing practice and sponge harvesting tools, and a group of coral reefs in its different forms. The citadel currently has been emptied of all the exhibits and antiquities. They are currently being held in a safe place because of the Syrian war, and many halls have been closed to the visitors. The citadel is still open to visit six days a week; however, it is closed on Fridays.

② Arwad Tower

The defensive maritime tower is situated on the eastern side of Arwad Island, which is situated directly on the port of Arwad, it has a square shape and an inner courtyard with two round northeastern and southwestern corners of fortified walls built of

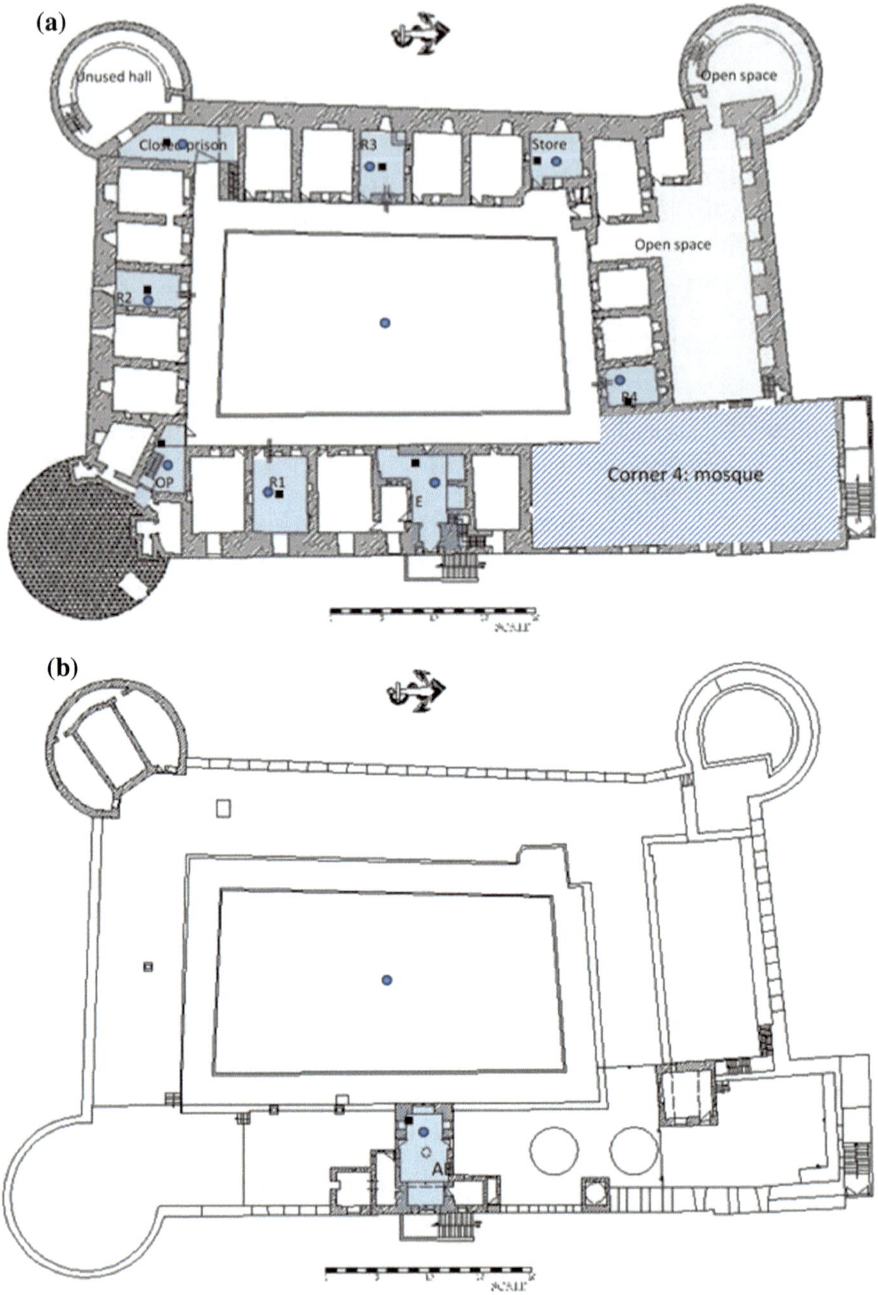

Fig. 4.5 Arwad citadel design and equipment locations. **a** Ground floor plan. **b** First floor plan. **c** Eastern façade. **d** Inner courtyard. *Source* Author based on the schemes of the Antiquities and Museums Department in Tartous

(c)

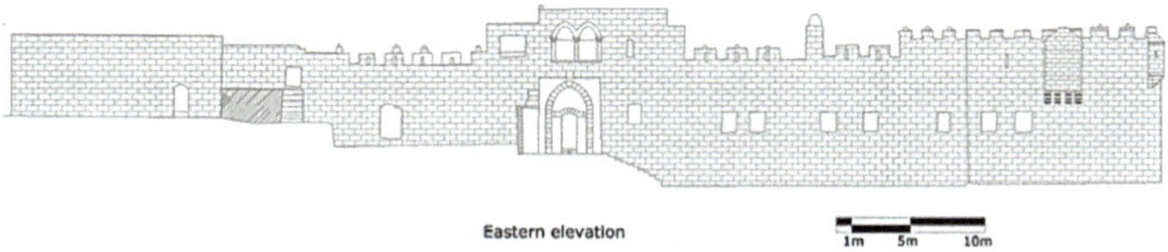

(d)

●Air flow anemometer ■ Temperature and relative humidity data logger

Fig. 4.5 (continued)

sandstone ashlar. The eastern facade is completely closed, while the western facade has the single main gate with a place above to put the flag or torch in it. The gate leads to the entrance of the tower which has a vaulted ceiling and contains two doors, one from the right side which leads to a small room and the other is directly opposite to the main gate leading to an open yard surrounded by fifteen halls including the entrance hall on the ground floor. The ceilings of the rooms are built in a vaulted style, and the rooms contain fireplaces for heating that are built within the walls. Only one room contains a well to collect water from the roof with an open hole in the roof above, and it is used when needed by the residents of the tower. All rooms on the ground floor are closed to the outside and opened to the inside, except two rooms of the corners of the tower from the west side. The first floor is accessed from a narrow stone stairway in the entrance, composed of five halls on the west side of the tower. The main hall which runs down the middle opens to the outside Iwan with a cross-vaulted roof above the entrance and leads to two rooms with dome-shaped ceilings on each side that characterize the architecture during the Ayyubid period. The roof is provided with a defensive rampart with a height of approximately 4 m to increase the fortification and equipped with defensive openings for control, archery, and dropping heat oil and stone gears from the four sides. The General Directorate of Antiquities and Museums in Tartous restored the tower. The roofs were restored with lime and the floors poured with traditional mortar, and they prepared a study to reuse the buildings, but it stopped because of Syria war in 2011. All information about features, materials, and elements of the case study sites are demonstrated in Table 4.1 and Figs. 4.5 and 4.6. The overview of all the measured rooms in the citadel and tower is shown in Table 4.2.

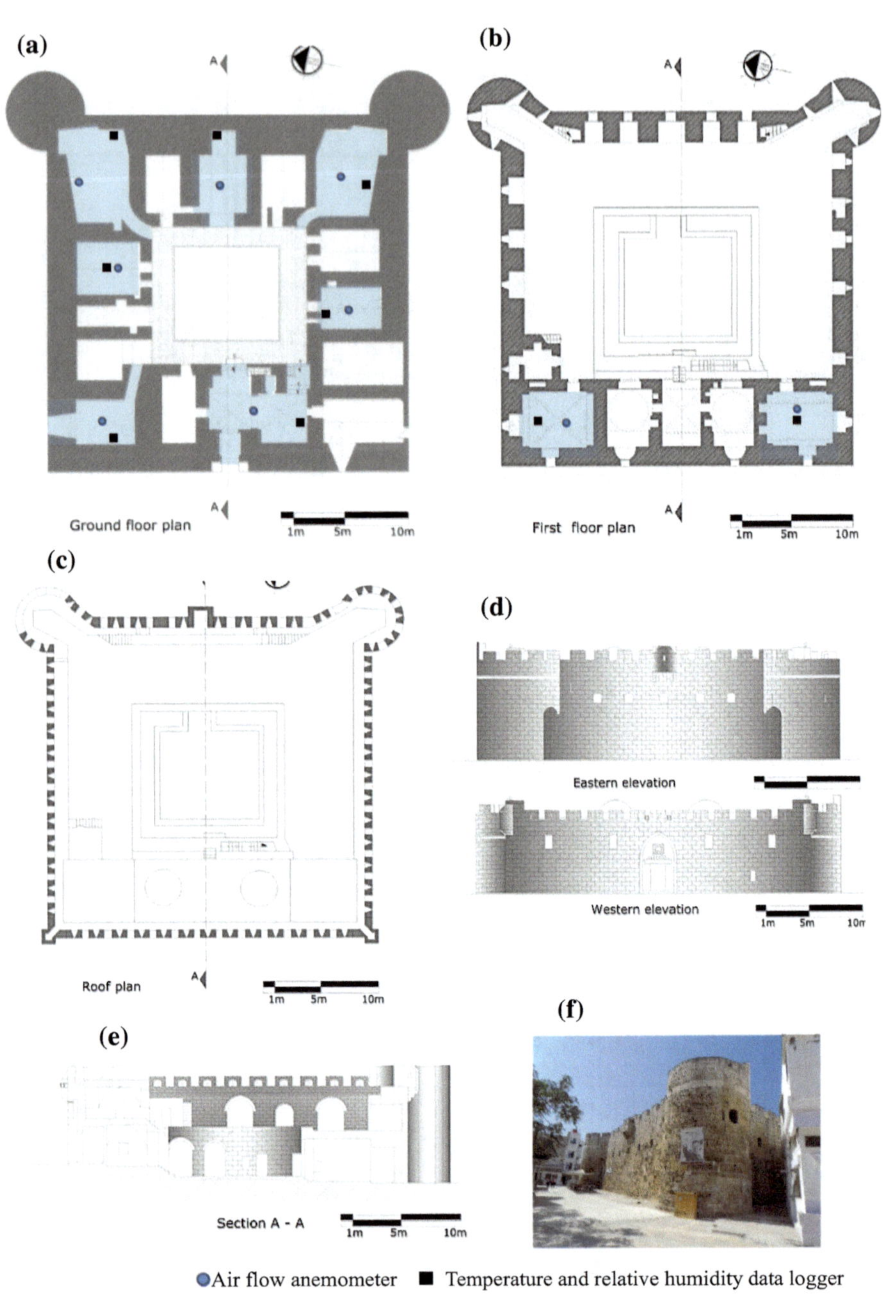

Fig. 4.6 Arwad tower design and equipment locations. **a** Ground floor plan. **b** First floor plan. **c** Roof plan. **d** Eastern and western facades. **e** Section A-A. **f** External eastern façade. *Source* **a**–**e** Antiquities and Museums Department in Tartous (2015)

Table 4.1 Materials and elements of the Arwad medieval forts

	Citadel	Tower
Location	The center of the island, 14 m above the sea level	The eastern side near the entrance of the island, 1 m above the sea level
Floors	Ground floor and partial first floor	
Shape and building orientation	Rectangular shape, the long axis runs from south to north, with three round towers and one rectangular. All of them are equipped with the opening of arrows, heated oil and stone gears	Almost square shape with two round northeastern and southwestern corners of fortified walls. The eastern facade is completely closed, while the western facade has the single gate
Material and construction	Vault—dome construction carried on massive sandstone load-bearing defensive walls	
Courtyard	Two inner courtyards. The main rectangular courtyard is (39.4 * 24.1) m, and the second one is (9.8 * 27.1) m.	One inner courtyard (11 * 12.1) m
Walls type	Massive sandstone masonry walls of double-layer	
External walls thickness (m)	2.3–2.7	2.15–3.1
External walls high (m)	14	8–10 (because the different ground level of the facades)
Courtyard walls (m)	0.85–1	
Internal walls (m)	0.65–1	0.85–1.3
Windows type	Single pane of glass and wooden frame with thin iron bars and without shutters	
Windows shape	The external windows with the arch ending and internal openings with a rectangular shape	Simple rectangular openings
Exterior windows' design	One or two casements can wide open, but they leave closed because of the privacy of the residential neighborhood	
Interior windows' design	One awning can just open with angle of 45 away from the wall that relieves the ventilation	
Exterior doors	confined to the gateway were made out of two layers of wood planks alternating between vertical and horizontal, the two layers hold together by iron studs, and reinforced and strengthened by metal plates	
Interior doors	The internal doors made from wood and almost the same size	
Insulations	No insulations, just the massive mass of the building itself in ancient time. After restoration, traditional white lime plaster for some internal walls and ceilings, and lime-concrete for floors	
Ancient systems	Some halls have fireplaces are built within the walls and wells dug into the rock	
Energy upgrades	No energy upgrades have been installed since the year of construction, except a small heating and cooling devices for the employees in the main office and guard room in winter and summer	

4.1.2 Zoom in Approach and the Microclimate Survey

This approach used the bottom-up monitoring and preserving processing upward and controlled from the low levels, where the information and decisions are informed by occupants survey and realistic discussions starting from small scope with all the needed details and incoming data from the environment to form a perception, then it grows eventually in complexity and completeness to meet local optimization and form the wide vision. The first three steps (define-measure-analyze) of the Six Sigma Circle (DMAIC/DMADV) were used in the Zoom in approach to provide the basic data for further research and steps of design, control, and verification using the simulation programs, as shown in Fig. 4.7. This survey begun with exploratory research, data collection of the island history, and climate studies to form the basic facts for the study. Then, it was followed by field measurements to quantify the

Table 4.2 Overview of all the measured rooms in the citadel and tower

Building	Floor	Room name	Room area (m^2)	Room internal orientation	Openings							Fireplace	Notes	Pictures
					Internal façade openings				External façade openings					
					Ordinary window	Ipper window	Door	Openings to wall ratio (OWR)	Window	Door	Openings to wall ratio (OWR)			
Citadel	F0	Room3 (C-F0-R3)	28.4	East	1	1	1	22%	-	-	-	1	The room contains a well to collect rainwater	
		Room2 (C-F0-R2)	23.4	North	1	1	1	21%	1	-	4%	1	External façade window is closed most time	
		Room1 (C-F0-R1)	41.8	West	1	1	1	15%	1	-	13%	1		
		Room4 (C-F0-R4)	19.6	South	1	1	1	29%	-	-	-	1	The room has an electric fan	
		Closed prison (C-F0-CP)	37.6	East	-	-	1	100%	-	-	-	-	Internal façade is too small and the room has	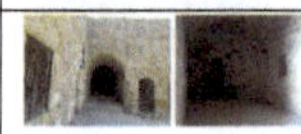
		Open prison (C-F0-OP)	22.8	North	-	-	1	100%	-	-	-	-	big discharged door of iron bars	
		Entrance (C-F0-E)	46.3	West	1	1	1	11%	-	1	32%	-	The entrance has big gate and the courtyart door of iron bars	
		Store (C-F0-S)	23.7	East	1	-	1	17%	1	-	3%	-	External façade window is closed with stones	
	F1	Coastguard room above the entrance (C-F1- AE)	39.3	West	2	-	-	7%	East (2)	-	16%	-	The openings are from the fourth sides	
									North (2)	1	9%			
									South (1)	1	11%			
Tower	F0	Entrance (T-F0-E)	44.7	East	1	-	2	47%	-	1	14%	-	Internal façade openings are quite opened without shutters	
		Opposite of entrance (T-F0-OE)	21.4	West	-	-	1	56%	-	-	-	-	An open Iwan	
		South circular corner (T-F0-SCC	35.2	West	-	1	1	34%	-	-	-	-	Internal façade is too small, no external openings	
		North circular corner (T-F0-NCC)	38.9	South	-	1	1	42%	-	-	-	-		
		North angle corner (T-F0-NAC	22.6	East	-	-	1	29%	1	-	6%	-	Internal façade is too small	

(continued)

Table 4.2 (continued)

		Guard room (T-F0-G)	21.3	South	1	1	1	16%	-	-	-	1	Internal façade windows are closed most time	
		Opposite of guard room (T-F0-OG)	16.5	North	1	1	1	27%	-	-	-	1		
	F1	South angle corner (T-F1-SAC)	28.2	East	-	-	1	9%	North (1)	-	4%	1	The room with a small hole in the ceiling	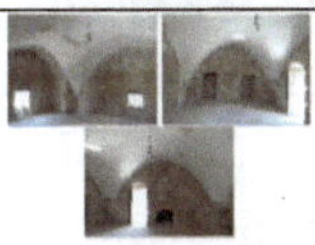
									West (1)	-	3%			
		North angle corner	30	East	-	-	1	11%	West (1)	-	3%	-		
		(T-F1-SAC)							South (1)	-	4%			

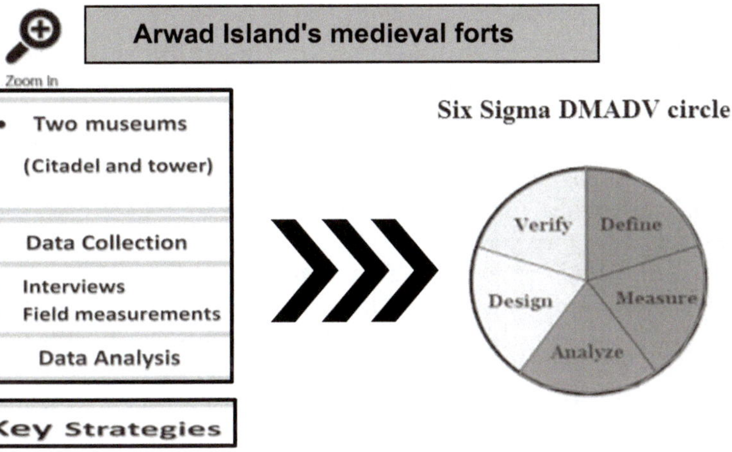

Fig. 4.7 Methodology structure

problem and to uncover facts and the actual conditions in the buildings. The study was summed up by an analytical approach in gathering the data and evaluating the indoor thermal environment for the studied buildings to find out the future actions needed for conservation.

① Data Collection

For this case study sites, three data collection and qualitative documentation methods have been used. Firstly, the review of the formal documents from the Department of Antiquities and Museums in Tartous, the related previous studies, field observation and non-structured interviews with the local people, employees and decision-makers about the island heritage and selected buildings (20 respondents), refer Appendix B. Secondly, the documentation of the selected buildings drawings, structure and materials, and the selected rooms information and notes, together with photo documentation and video filming. Thirdly, the field measurement using the equipment provided by the University of Chongqing to carry out measurements and those that could be entered into the case study site, the main thermal parameters (air temperature, relative humidity, and airflow) have been measured.

(1) **Field Measurement and Procedures**

For determining the indoor thermal environment for buildings, the monitoring period was during the hottest month, for continuously 13 days from 19th to 31st August 2015. The main thermal parameters have been measured with the help of the meteorologist's station to get the rest of the information and weather conditions data, and the field measurement of each room was taking place for three continuous days, (only one person is needed to complete the data collection in each building, and two persons for manual outdoor and indoor wind flow records). Measurement equipment includes four temperature and relative humidity data loggers, and two Testo 405-v1 anemometers for measuring the airflow and temperature. During the measurements, the most rooms of the two buildings were closed for peoples' visit because of the current country war and the safety of buildings and archaeological collections inside the citadel. The thermal recording devices that used to monitor indoor air temperature and relative humidity were placed according to each room's conditions and furniture at a height of 1.1 m above the floor, and they were set to make the thermal measurements during the period of 10 min intervals and recorded continuously for three days in each room, as guided by the relevant standard (ASHRAE, ANSI 2004).

The simultaneous measurements of the indoor and outdoor temperature and airflow were taken by two devices, one for indoor records and another for outdoor records. The readings of the measurements have been registered at the same time using cell phone connection between the author and her assistant. The internal records of simultaneous measurements were taking place at the center of each selected room of both historical buildings with a sensor directed toward the openings that open to courtyards. The external measurements of the tower were gauged from a location in the north breakwater of Arwad Island. The external measurements of the citadel were gauged from a point in the center of the courtyard. This location was chosen because the width of the citadel's courtyard provides much of the ventilation for the rooms and using another location was not feasible due to the topography and the neighboring buildings.

Indoor and outdoor airflow and temperature measurements were recorded at the same time as the recording of the indoor temperature and humidity. The simultaneous measurements were continued for three days and recorded three times each day, each recording took about three minutes (ASHRAE, ANSI 2004). The largest and smallest values were recorded, and the arithmetic mean and the difference between outdoor and indoor records have been calculated for each time as a first stage, then for each day, to have the daily average of indoor and outdoor records in August with its standard deviation for each room.

② **Data Analysis**

After the completion of measurements, the data were exported to Microsoft Excel 2010 spreadsheet for analysis. All data collection and the physical measurements of the indoor environment were analyzed and then compared with previous studies and standards to get into treatment and control strategies.

4.1.3 The Climate and Outdoor Weather During the Period of Measurement

Heat and humidity are the main issues considered for achieving comfort in the climate of Arwad Island. Arwad Island's climate is generally moderate Mediterranean in nature. Latitude of 34.53°N and longitude of 35.53°E provide a humid environment in summer and rainy winter weather with south to southwestern prevailing winds (Amer and Abou Hamed 2011). The average value of the diurnal highest temperature in Tartous is in August at approximately 30 °C, and the mean value of the minimum temperature is in the winter in January at approximately 7 °C. The average humidity is between 60 and 75% (Rahman and Rahman 2015). The temperature in this region has increased by 1.1 °C within the last 60 years (Ricciardia et al. 2016; Rahmoun et al. 2016). The monitoring period was performed during the hottest month, for 13 consecutive days from the 19th to the 31st of August 2015. During this period, the prevailing winds were westerly winds; wind velocity was approximately 1.3–3.1 m/s in the daytime; the highest temperature was 36 °C; and the lowest temperature was 27 °C. The relative humidity in the atmosphere was very high with mean values ranging between 70 and 90%. The island experienced 13 h of daylight per day and 11 h of sunshine per day, 79% of daylight hours, and 28 °C sea temperature in August in Tartous (Rahmoun et al. 2016).

4.2 The Current Microclimate Situation of Arwad Island's Medieval Forts

4.2.1 Field Measurement Results

① **Indoor Air Temperature**

The average air temperatures in the citadel were between 28.8 °C, during nighttime, and 34.2 °C, during daytime, while the range in the average air temperature in the tower was between 27.3 and 33.3 °C, Fig. 4.8. The first floor rooms in the tower have a higher temperature than the ground floor rooms, as the air temperature in those spaces is easily influenced by the outdoor air temperature with the relatively large amount of solar radiation from the openings. Furthermore, they are influenced by the

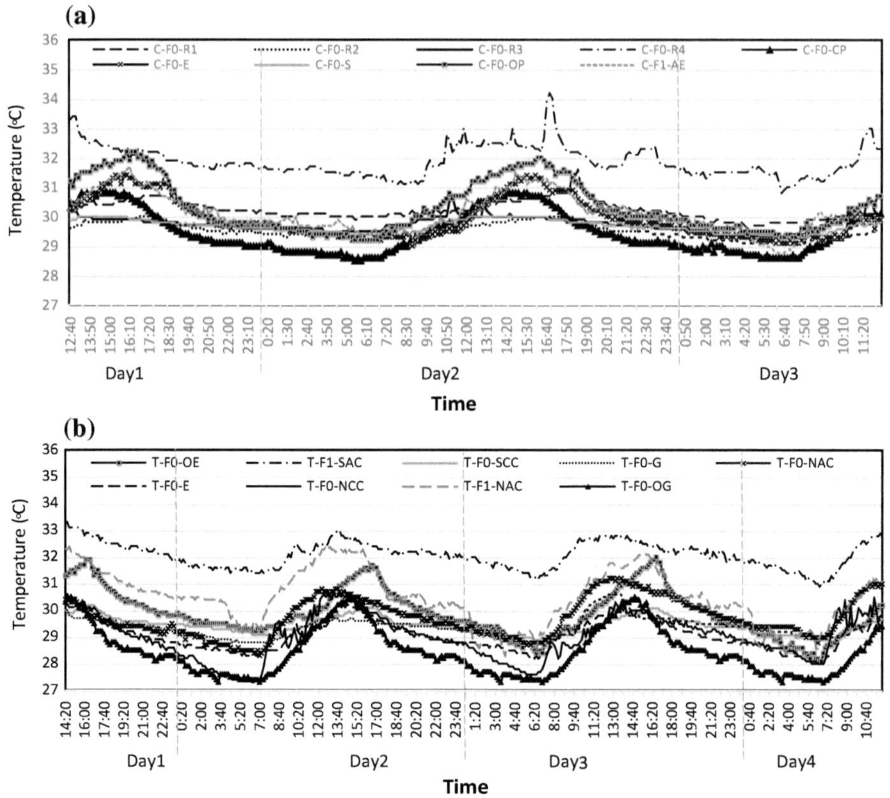

Fig. 4.8 Measurement results of the indoor air temperature. **a** Citadel rooms. **b** Tower rooms

heat from the roof and walls during the daytime. The orientation of the rooms of these two building played a significant role in increasing the air temperature. The rooms of the internal facade that are oriented to the south and west in both buildings have the highest indoor temperature. While the lowest indoor temperature was measured in the rooms of the internal facade that are oriented to the north and stay shaded during daylight hours, followed by those that are oriented to the east. Citadel rooms have more thermal stability between night and daytime than tower rooms. While the corner rooms with their big open doors (OP and CP) of the citadel were noticed to have obvious differences in temperature of approximately 3 °C between night and daytime and had the lowest temperature at night (28.8 °C) compared to other rooms. Most of the rooms' temperatures in the tower vary between night and day by approximately 3 °C.

② Relative Humidity

The relative humidity was between 56 and 82.5% in the citadel and between 50.3 and 90% in the tower, Fig. 4.9. The lowest relative humidity values were recorded in

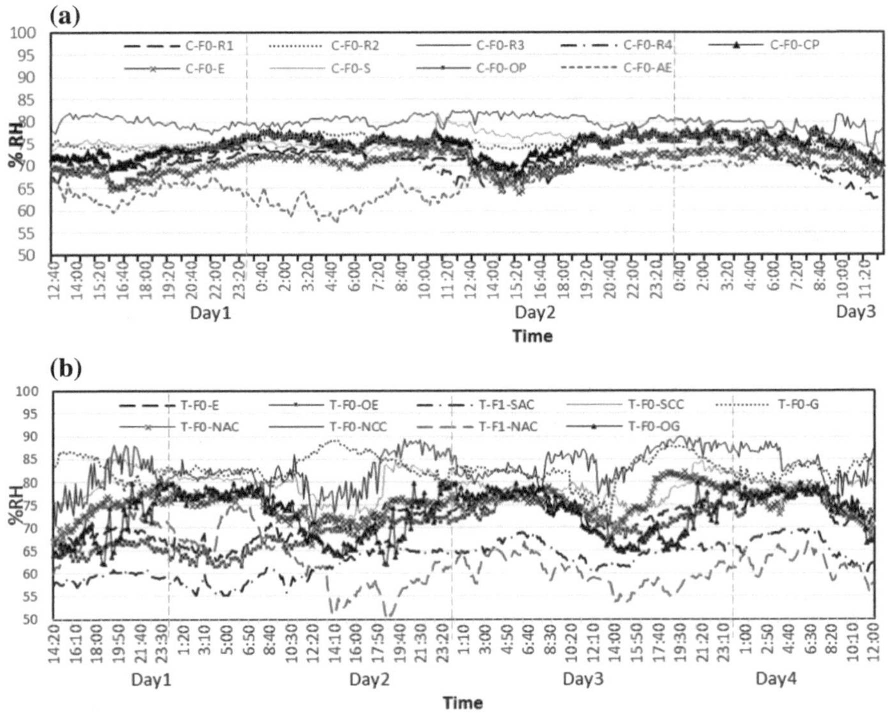

Fig. 4.9 Measurement results of the relative humidity. **a** Citadel rooms. **b** Tower rooms

the first floor rooms (AE, NAC, and SAC), and in the rooms that are oriented to the west in the citadel and tower. The highest relative humidity values were recorded in the rooms oriented to the east and the north (R3, Store, and R2) of the citadel and in the circular corners (NCC and SCC) and rooms oriented to the south (G) in the tower. The lack of ventilation and the effect of moisture absorption of the sandstone may perform the role that harms building materials and causes deterioration of the built environment in these historic buildings. Most of the measured temperature and relative humidity values exceeded the relevant standard for sedentary activity during summer (Erhardt and Mecklenburg 1994). This can damage the building structure and materials, cause human discomfort, and promote mold growth (Lourenco et al. 2006; Sileo et al. 2017). High humidity typically occurs at night and early morning. The castle rooms have a relatively stable humidity in nighttime and daytime, while the closed exterior facade and small openings toward the courtyard of the tower contribute to a significant increase in moisture.

The daily average of the indoor air temperature and relative humidity in the measured rooms of the selected buildings are also calculated and configured as shown in Fig. 4.10.

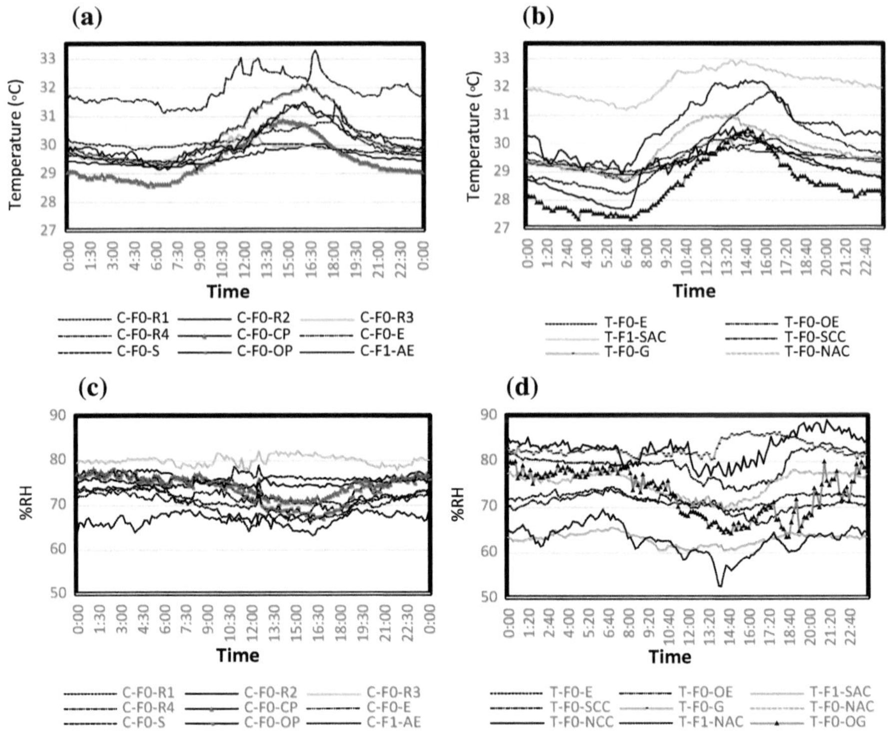

Fig. 4.10 Daily average of the measured parameters. **a** The indoor air temperature of the citadel rooms. **b** The indoor air temperature of the tower rooms. **c** The indoor relative humidity of the citadel rooms. **d** The indoor relative humidity of the tower rooms

③ The Indoor and Outdoor Wind Flow

The highest value of ventilation exists in the open prison (0.24 m/s) in the citadel and in the entrance (0.17 m/s) and the Iwan (OP) (0.12 m/s) in the tower, as shown in Fig. 4.11, because they are entirely open to the inner courtyard. The rooms of the first floor in both buildings have relatively high ventilation rates with a small difference between outdoor and indoor wind flow because of their openness to the wind direction and to the outside. The size of the courtyard in these two buildings did not play a significant role in providing ventilation for the rooms. The ventilation still has very low values in most rooms ranging between 0.05 and 0.08 m/s. This is due to the enveloping massive walls closing toward the outside in the original design and the partial opening of the internal windows added later during operation, which further minimizes the ventilation. Most of the time, the outside windows of the citadel are closed to offer privacy to the residential neighborhood of the citadel. Both buildings need an increase in ventilation ranging between 0.06 and 0.1 m/s compared to the 0.15 m/s of the ANSI/ASHRAE standard for occupant comfort and collection conservation (ASHRAE, ANSI 1992).

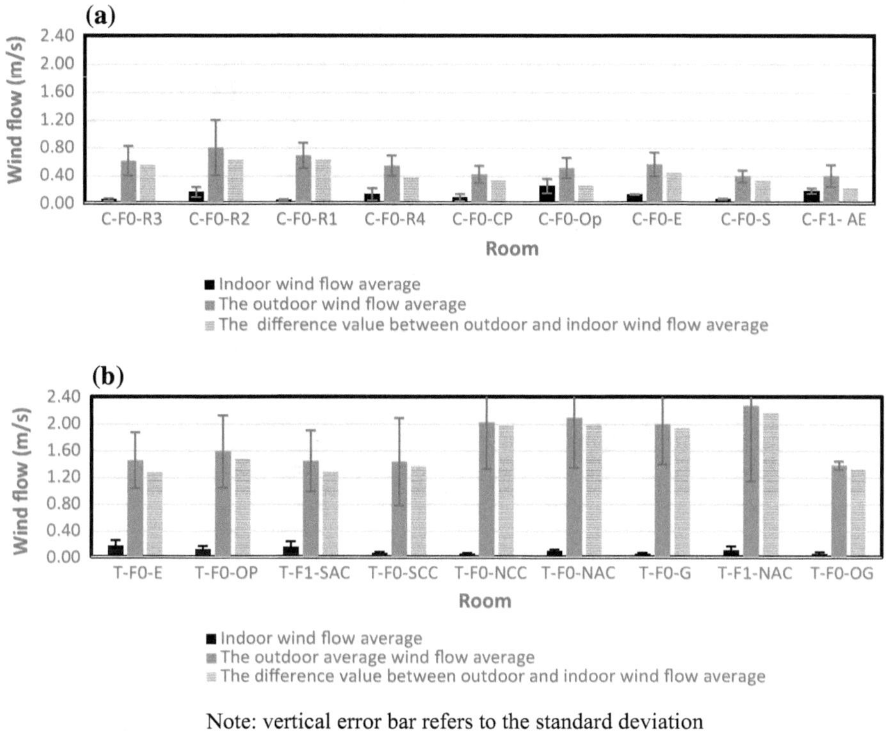

Fig. 4.11 Daily average of indoor, outdoor and the difference between indoor and outdoor wind flow during the measurement period. **a** Citadel rooms. **b** Tower rooms

4.2.2 Interview and Field Observation

① Interview Results

Twenty unstructured and exploratory interviews with local people and the director and employees of the selected buildings about the island heritage and their buildings were conducted at the time the physical environmental measurements were taken. In general, Arwad Island's buildings suffer from the durability of the material against the humidity and salinity of the sea. Heritage considerations in the island are limited to the local guide of using sandstone in the exterior siding in order to give the traditional character of the historical buildings on the island, but the sandstone outer skin did not bear the humidity well and some stones have fallen down. The staff members of the citadel agreed that the location of their offices, which are oriented to the south, maintain higher temperatures than rooms with a northerly position and are uncomfortable in summer. They state that the material of restoration should have more resistance to the humidity. Since a short time after restoration, the walls and the mortar do not appear to have been refurbished. Visible damage of the building

Fig. 4.12 Visible damage of building material. **a** Green stains on the wall. **b** Damp and mold on the wall

materials is presented by the green stains and mold on the walls and can be seen in many rooms in both the citadel and tower, especially on the ground floor, due to the sandstone's humidity sensitivity, Fig. 4.12. There is no documentation of the original design and traditional type of window frame and model. Moreover, the current design of the windows does not provide proper ventilation during the summer. No energy upgrading inside the museum rooms or microclimate study has been conducted.

② Mold Growth, Visitors' Comfort, and Preserving Collections

A moisture and mold growth problem occurred in the indoor rooms of the analyzed building. Solar radiation overheats the roof and massive walls, which then store and transmit the heat to the indoor environment. This forms an uprising current of overheated air in summer and sucks in outside air from the window openings. The external air entering through the awning windows is not undisturbed free air but is the overheated air of the internal boundary layer formed on the walls. The moisture content increases in the rooms because the moisture in the walls hit by sunshine evaporates and is trapped in the uprising flow along the wall surface, causing the growth of mold. Therefore, there is a need to secure ventilation and reduce the moisture. The various types of sandstone on the coast are affected by environmental conditions with the possibility of water leakage inside them. Hence, it is necessary to evaluate the durability and moisture susceptibility of building materials for the selection and upgrade of conservation materials. To affirm the results, particularly during the opening times of the museum, a simultaneous measurement of the indoor and outdoor temperature was done for each room between 9 am and 5 pm each day. The lowest indoor temperature is in the rooms with the internal facade oriented to

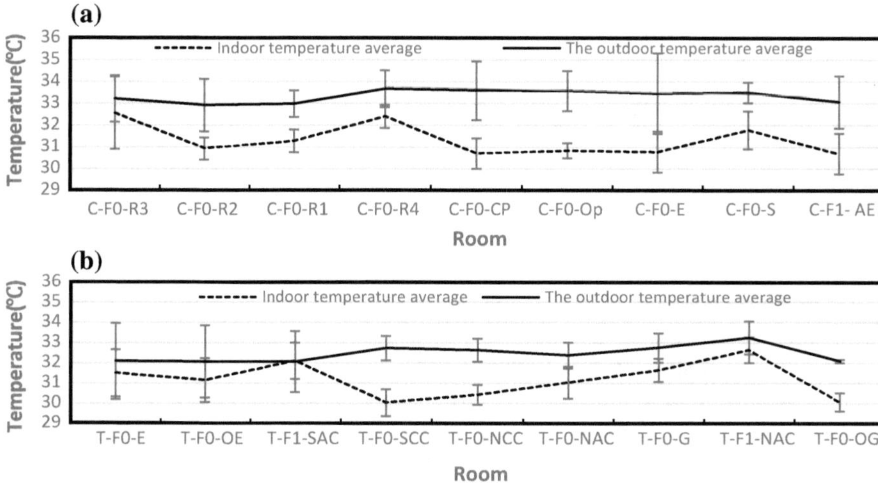

Note: vertical error bar refers to the standard deviation

Fig. 4.13 Daily average of indoor, outdoor and the difference of the indoor and outdoor temperature during the measurement period. **a** Citadel rooms. **b** Tower rooms

the north (R2 and OG) and in the corner rooms of the ground floor (CP, SCC, and NCC). The results reveal the general increase in the temperature as high as 36 °C at approximately noon demonstrates the importance of considering noon and early afternoon hours as the times that the buildings receive the most heat during the day. The high temperature occurred in the rooms of the internal facade, which are oriented more to the south (R4 and G) compared to the remainder of the rooms, followed by the rooms oriented to the east during the opening times in both case study sites, as shown in Fig. 4.13. The first floor rooms in the tower have the highest temperature compared to the remainder of the tower rooms, while the first floor room in the citadel has a relatively low temperature compared to the remainder of the citadel rooms. In the comparison of the rooms on the first floor in the citadel and the tower, the results yielded the latter has a higher indoor temperature by two degrees, and the indoor temperature in them is similar to the outdoor temperature and is sometimes higher, especially at noon. Perhaps the reason is due to the large space that is open to the outside with many openings compared to the tower rooms with smaller spaces. All of the above emphasize the importance of accounting for the microclimate features during the operating period in the analyzed buildings' rooms to secure visitors comfort and to preserve collections.

4.3 Key Strategies and Recommendations

The following control strategies during the hot period would be the basis to support further experimental and numerical analysis of natural ventilation and thermal conditions on Arwad Island and would be useful for preserving the buildings and respect the characteristics of the local environment:

① Passive ventilation using natural and variable wind speed secures moving ventilation from the internal courtyards to the rooms and helps maintain comfortable conditions inside the buildings to endure the annoying moist heat in the summer with the lowest costs. Redesigning the window types and their open-close mechanisms will help decrease moisture and improve the environment, controlling the prevailing winds in order to provide temporal ventilation and appropriate airspeed. The windows located opposite each other in the citadel's rooms are useful for cross-ventilation. The openings at a high level would naturally aid in reducing the moisture and would vent the hot air out when opened.

② Consideration should be taken to preserve collections and provide the visitors comfort. The mechanical fan ventilation systems could be a solution to remove larger quantities of moisture quickly from the tower's ground floor rooms to the outdoors because it is closed to the outside and not receiving airflow through the openings from the small internal courtyard. If a HVAC system should be needed to have constant thermal parameters and stability in condition after investigating the criteria, one or two rooms in each building could be treated to meet the standards for conservation in accordance with the specifications of the museum community.

③ Consideration should be taken to organize the space in regards to its use depending on the time of day and the orientation. Suitable places to properly and safely display the exhibitions should account for the lighting system, the need for shading, and thermal insulation to reduce the heat flow from the sun as it reflects against the walls and roof. Special attention should be given to the heat from the roofs because they receive the largest amount of solar radiation. Investigating the properties and making a microscopic analysis of the sandstone is imperative and recommended to understand the performance and moisture control of the stone. Strengthening the natural elements such as heat, moisture, insolation, and wind is important to ensure the proper thermal requirements and the balance between the effects of the external environment and the internal activities of the occupants and visitors inside the museum during different climatic periods.

Future recommended studies of this step should include a detailed examination of parameters affecting the built environment through a computer simulation analysis and accompanied by prototype construction for measurements to provide a comprehensive study, especially in validating theoretical methods for thermal predictions of any building design. Further studies concerning reducing moisture and increasing thermal comfort might be performed in residential buildings to have a broader image of the thermal comfort phenomenon on Arwad Island and in the old city of Tartous.

4.4 Summary

The study of this chapter documented and identified the existing situation of the selected Arwad medieval forts' structures. It determined the current thermal conditions and their effect on the comfort of the occupants, building material, and the condition of the museum's collections. It also provided sets of data of microclimate features of the heritage structures and discovered appropriate strategies to control and achieve climate control. By analyzing the results obtained in the historic buildings, most of the measured temperature and relative humidity values exceeded the relevant standard for sedentary activity during summer conditions, which can damage the building structure and materials, cause human discomfort, and promote mold growth. The case study sites in Arwad Island have an undesired excess of dampness after a daily increase in air temperature during the summer. Additionally, the current openings affect the airflow inside the building and cause overheating. Measurement results confirmed the importance of providing cross-ventilation and the need for reducing the relative humidity during the summer. The control strategies during the hot period are also discussed, and further studies were recommended. As the first microclimate research on Arwad Island, the findings are important to explore the knowledge of the current thermal situation of the local historical buildings and determine the problems in order to generate future guidelines for maintenance, restoration, and operation.

References

Amer, R., & Abou Hamed, F. (2011) Optimal study design for Wind farm in Arwad Island. *Energy Procedia, 6*, 721–735.

Antiquities and Museums Department in Tartous. (2015). *Arwad Island's historic buildings*. Tartous, Syria: Antiquities and Museums Department, Historic Buildings Section.

ASHRAE. (2003). *2003 ASHRAE handbook: heating, ventilating, and air-conditioning applications*. Refrigerating and Air, Atlanta, GA: American Society of Heating.

ASHRAE, ANSI. (1992). *Standard 55-1992, thermal environmental conditions for human occupancy*. Atlanta, GA: American Society of Heating, Refrigerating and Air-Conditioning Engineers.

ASHRAE, ANSI. (2004). *Standard 55-2004, thermal environmental conditions for human occupancy*. Atlanta, GA: American Society of Heating, Refrigerating and Air-Conditioning Engineering.

Balocco, C., & Calzolari, R. (2008). Natural light design for an ancient building: a case study. *Journal of cultural heritage, 9*(2), 172–178.

Dumper, M., & Stanley, B. E. (2007). *Cities of the Middle East and North Africa: A historical encyclopedia*(p. 105). Santa Barbara, CA: ABC-CLIO Inc.

Erhardt, D., & Mecklenburg, M. (1994). Relative humidity re-examined. *Studies in Conservation, 39*(Supplement 2), 32–38. In *Preprints of the Contributions to the Ottawa Congress, 12–16 September 1994. Preventive Conservation: Practice, Theory and Research.*

Hassan, M., & Rahmoun, T. (2005). *Future planning for marine tourism in Tartous: Studying nautical tourist path linking among four important case study sites: Antaradus a tourist trade project, archaeological Arwad Island, ALhbas Island, archaeological area Amrit*. Lattakia, Syria: Graduation's Thesis, Faculty of Architectural Engineering, Tishreen University.

Land Identification and Registration Committee. (2014). *Demographic and real estate statistics of Arwad island*. Tartous: Tartous Bureau of Statistics and Municipality of Arwad island.

Lourenco, P. B, Luso, E., & Almeida, M. G. (2006) Defects and moisture problems in buildings from historical city centres: A case study in Portugal. *Building and Environment, 41*, 223–234.

Nicolle, D., & Hook, A. (2004). *Crusader castles in the Holy Land, 1097–1192*. Oxford: Osprey Publishing Ltd.

Rahman, S., & Rahman, M. A. (2015). Climate extremes and challenges to infrastructure development in coastal cities in Bangladesh. *Weather and Climate Extremes, 7*, 84–95.

Rahmoun, T., Hassan, M., & Alhasan, W. (2016). Protection strategy for the coastal areas of climate change "Syrian Coast, Tartous City". *Research Journal of Applied Sciences, Engineering and Technology, 12*(3), 264–271. ISSN 2040-7459.

Ricciardia, P., Zilettib, A, & Burattic, C. (2016). Evaluation of thermal comfort in an historical Italian opera theatre by the calculation of the neutral comfort temperature. *Building and Environment, 102*, 116–127.

Sileo, M., Fabrizio, T. G., & Masini, N. (2017). Low cost monitoring approach for the conservation of frescoes: The crypt of St. Francesco d'Assisi in Irsina (Basilicata, Southern Italy). *Journal of Cultural Heritage, 23*, 89–99.

UNESCO. (2016). *L'île d'Arwad*. Retrieved from http://whc.unesco.org/en/tentativelists/1303/. Accessed September 12, 2016.

Živković, V., & Džikić, V. (2015). Return to basics—environmental management for museum collections and historic houses. *Energy and Buildings, 95*, 116–123.

Chapter 5
Microclimate Field Survey in the Old City of Tartous, Syria

Monitoring heritage buildings and sites is an essential part of preserving their architectural, artistic, historic, and social values and developing possible techniques to save their structures and environments. The prediction of potential risks is vital to consider the appropriate actions for preserving heritage buildings and sites through different conservation processes and interventions. Conserving the cultural heritage can extend the life of these valuable resources of past cultures and improve the economic conditions of a society. After the latest serious heat waves observed in Syria and the expected rise in temperature in the future due to climate change impact, which would increase in the number of air-conditioning systems and create extra loads during the hot period, the local calls in parallel with the global one called to minimize the financial and ecological expenses of energy consumption and encourage the energy-saving measures and passive cooling strategies by using the natural ventilation to raise the indoor airflow and enhancing the thermal comfort without having any adverse effects on the occupants' comfort and life quality. Syria does not have adaptive environmental comfort standards. There are no field study data on the prevailing thermal conditions or occupant perceptions and preferences in the historic buildings, dwellings, and public buildings of Syria. Furthermore, there is no assessment approach for evaluating the performance of the buildings after completion/refurbishment and their use/reuse. Based on the above, this field study was performed and covered the old city of Tartous in Syria during the summer of 2016 to gather microclimatic data and determine the thermal comfort range of users of historic buildings in Tartous through thermal surveys and field measurements in order to perform a comparative analysis with the existing international standards of the indoor environment.

5.1 Overview of the Case Study Sites

Tartous city has a moderate Mediterranean climate; it has a humid environment in summer and rainy winter weather with south and southwestern prevailing winds (Amer and Abou Hamed 2011). Heat and humidity are the main issues to be considered for achieving comfort in this climate (Hassan et al. 2018). The monitoring

M. Hassan and H. Xie, *Climate Change and Conservation of Coastal Built Heritage*, https://doi.org/10.1007/978-981-13-8672-5_5

period for each place was performed during the hottest month, which was August 2016. During this period, the prevailing winds were westerly winds; the wind velocity was approximately 0.9–4.5 m/s in the daytime, the highest temperature was 35 °C, and the lowest temperature was 28 °C. The relative humidity in the atmosphere was very high, with mean values ranging between 57 and 77% (Meteorological Center of Tartous 2016).

5.1.1 Tartous Citadel Fabric and the Selected Historic Dwellings

Tartous Citadel on the Syrian coast was the headquarters of the Knights Templar, and the last crusader strongholds in the Levant mainland (Kennedy 1994). They built the Citadel in the seafront with double concentric walls and a double moat, and the famous Knights Templar buildings presented by the galleries, Banqueting Hall and the Chapel (Hassan and Rahmoun 2016). Now it is the historical center of Tartous city and well-known as the old city of Tartous. Due to the importance of the old city, it was registered as a national heritage with the decision of the Ministry of Culture No. 139/A dated 9/6/1999, and it has been submitted to the world heritage tentative list since 1999. Despite the first Phoenician presence in the city of Tartous, the basic core for the emergence of the old city of Tartous current form dates back to the Knights Templar. The main purpose of building the Tartous citadel was defensive and military purposes then it has been inhabited later in Muslim periods. The residents have demolished several parts of the defensive walls in order to build their houses. The civilian dwellings started grafting onto the ruins of both the inner wall and the outer turreted wall of the Templar fortress, the whole forming the oldest quarter of the city.

The transformations of the Knights Templar citadel during the Muslim period resulted in the birth of the traditional house of Tartous as an original type of urban environment, see Figs. 5.1 and 5.2. It is a house of simple plan which consists of three main elements: an entrance, a center (an inner courtyard or a central hall) and rooms of the house. These elements extend over one or two floors, over the ancient ruins of the Tartous Citadel. The traditional house of Tartous also distinguishes itself from other Muslim dwellings by its opening to the outside; its localization in the floor allowed it to have windows opening onto streets and places (Karfoul and Mouton 2008). These series of spontaneous interventions formed the construction of a residential city on the remains of a Crusader castle. The urban organisms have been constructed through time by the sum of diverse cultural influences, and they conserve within their urban fabric all the traces of their history and identity. The morphological processes that have affected the building fabric through time and determined its present structure, and gave it the morphological and spatial value (Neglia 2004). The reinforced concrete in thin layers was used in the upper part of

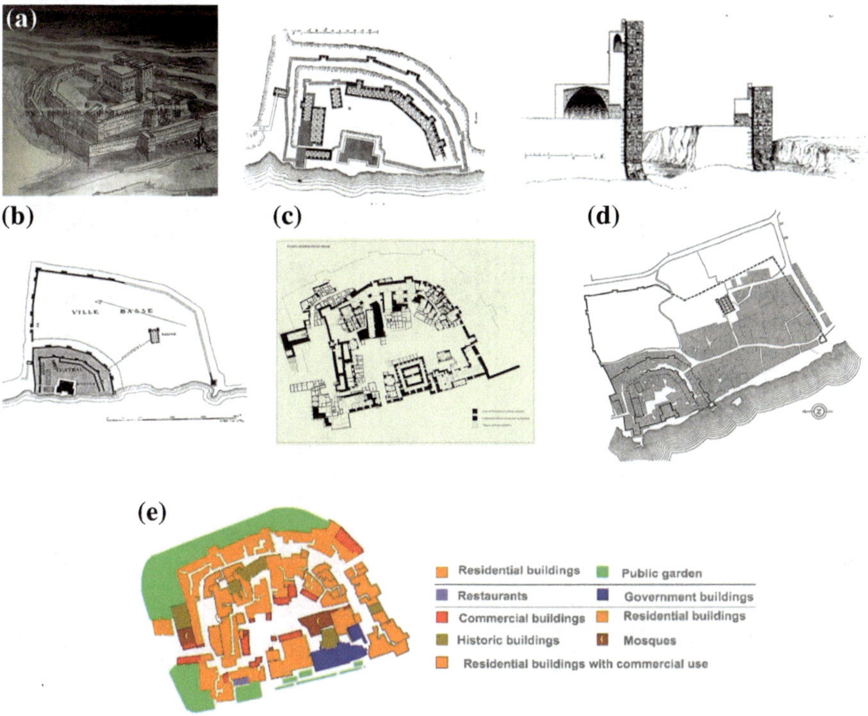

Fig. 5.1 Transformations of the Tartous Citadel: **a** the perspective, plan, and the section of inner and outer Crusader castle' walls in 1180 AD according to the Orientalist Rey, **b** Tortosa bishopric consisted of the castle and the cathedral of Our Lady of Tortosa and the within the third historic wall, **c** the Ottoman city, **d** the later city extension, **e** the current buildings usage of the old city of Tartous. *Source* (**a**) Rey (1866), (**b**) Pillet (1929), (**c**, **e**) Antiquities and Museums Department in Tatous (2015), (**d**) Deschamps (1973)

the Citadel that characterizes the architectural style in the late Ottoman period and the beginning of the French mandate period.

The Questionnaire survey randomly conducted in the urban fabric and residential buildings of the old city of Tartous, as shown in Fig. 5.3. The field measurements were carried out into three residential houses of Knights' Hall (Hall of the Crowns). It is located in the northern part of the castle and its dimensions 47.8 × 18.9 m, which is the only two-story building still standing from the crusader inner fort block. Nowadays, the Knights' Hall reveals phenomena of transformation from specialized to residential. The formation of residential fabric inside the Crusader hall has definitively changed the consolidated sense of space and the structures themselves since this took place by building over ancient structures and transforming the ruins of the gothic architecture into the residential fabric of the medieval Islamic city. The ground floor was divided into several sections. The outer section was a corridor with darts openings overlooking the circular wall. The interior section was used as a water

Fig. 5.2 The old city of Tartous **a** the surrounding moat, **b** the north side of the old city of Tartous, **c** the remaining of the city walls, **d** the old city square

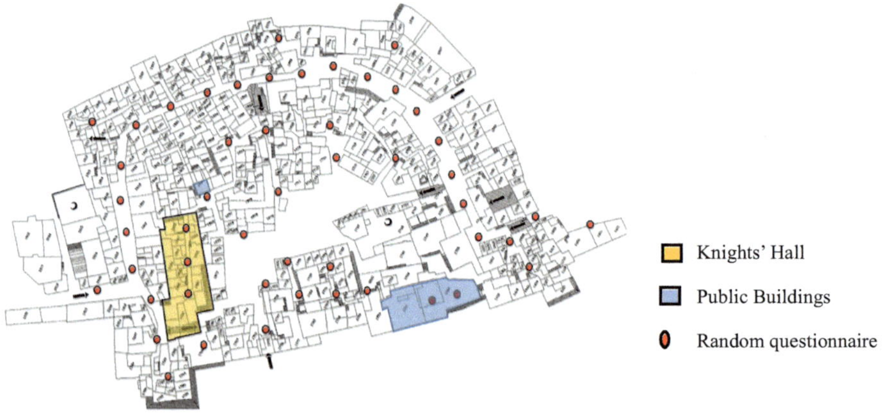

Fig. 5.3 Case study site of this chapter in the old city of Tartous

reservoir and a storage room. The first floor consists of a memorial hall with a cross-vaulted roof and crown columns, the hall was a beacon through windows from the four sides and from thc ceiling (Braune 1985). The collapse of the upper levels of the north wall and the columns supporting the cross-vaulting, which happened when the Muslims conquered the citadel, facilitated the construction inside the Hall, in which small dwellings were built later in the Ottoman period, aided by the permanence of the four perimeter walls, and along a street running down the middle of the Hall, where the original hall seemed to consist of two galleries. Some column crowns still exist inside the dwellings (Neglia 2009).

The selected three traditional houses are shown in Figs. 5.4, 5.5 and 5.6. They were selected for this survey based on the following selection criteria: the significant value within a consolidated urban and architectural context of the space on the first floor of the big hall, the important thermal mass, the use of natural or mechanical ventilation during the hot season, and the availability of people to participate in the investigation. These structures are built of sandstone. The maintenance of the buildings and their facades is not good in most cases, but generally, there are in good condition. The essential problem is the peeling paint of their internal walls and roof. The first house called Kabalan house (H1) distributed on three floors with openings to the outside from the south direction, the first floor consists of an entrance, a simple living room with kitchen and stairs. While two bedrooms distributed on the other two floors. The other two houses (called houses with a square hall) consist of three main elements: an entrance, a big living room that called square hall, and the rooms of the house. These elements extend over one floor, with a different orientation. The first one called Al-Moubaied house (H2) with south and west orientation, and the other one called Sanaa Shalak house (H3) with north and east orientation.

Fig. 5.4 The knights' hall: **a** the collapsed cross-vaulted roof of the knights' hall (**b**), **c** the residential buildings abutting the inner walls and crowns of the hall, **d** the southern facade of the knight's hall and the upper part of the southern facade of the knights' hall

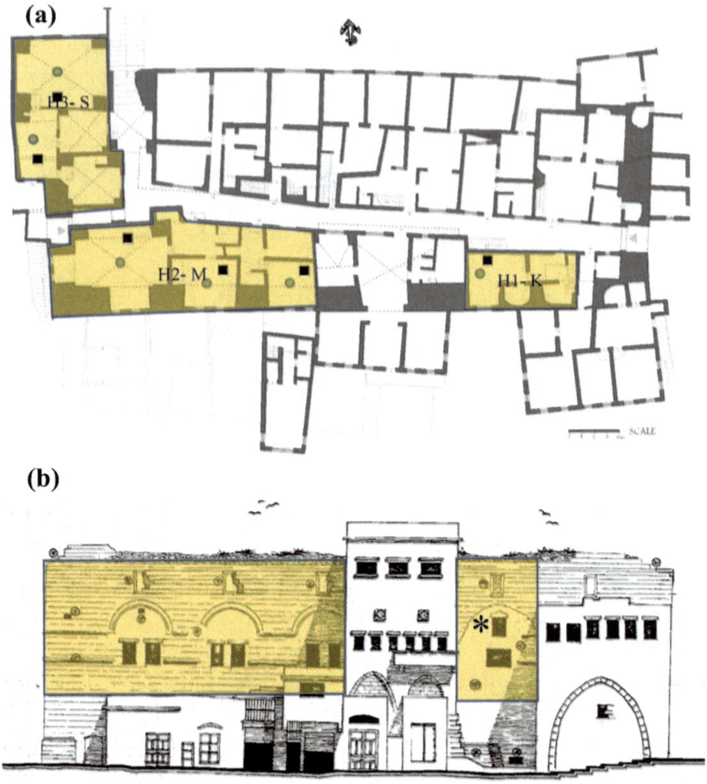

●Air flow anemometer ■ Temperature and relative humidity data logger

*Kabalan house is three level, the living room is shown in the plan of the first floor and the data logger in the bed room of second floor was hanged in the same place as in the first floor.

Fig. 5.5 Selected houses of the knights' hall and the equipment locations: **a** plan. **b** Façade. *Source* Author based on the schemes of the Technical Office of the old city of Tartous

5.1.2 Historic Public Buildings' Structure and Design

The survey was carried out in two historic office buildings, one public hall and one print shop, as shown in Figs. 5.7 and 5.8. The office buildings are located on the waterfront of the old city of Tartous and rise approximately 1 m above sea level. Both buildings share a cross-vaulted ground floor of massive sandstone masonry walls. The ground floor was built in the Crusader period, and it is currently used as a public multifunctional gallery (G). All selected public buildings have western orientation and are built of sandstone with different thicknesses influenced by the building materials and types when they were built. The first selected office building is the Department of Antiquities in Tartous (D). It was renovated in 2007, and the

Fig. 5.6 Selected houses of the knights' hall: **a** Kabalan house. **b** Al Moubayed house. **c** Sanaa Shalak house

distortions and additions were removed. The building has two floors, and the first floor contains two parts; the first part was built in the Ottoman period, and the other part has rooms that surround an inner yard and was built in the late Ottoman period. The second floor was built above the Ottoman part around the middle of the nineteenth century.

The second selected building is the Technical Office of the old city of Tartous (T), and it belongs to the municipality of Tartous. This building also has two floors. The first floor was built in the late Ottoman period as a courtyard house with cross-vaulted rooms; then, it changed in the early nineteenth century to be a central hall building

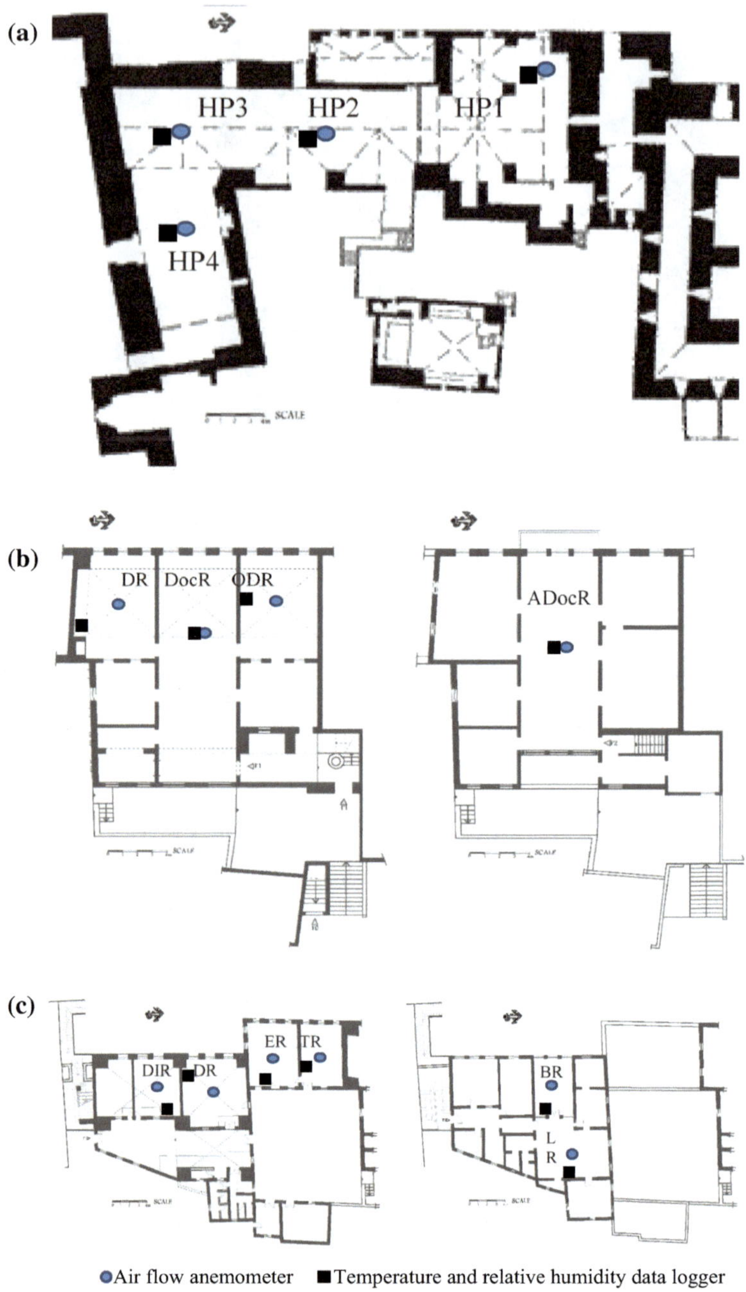

Fig. 5.7 Plans of the selected historic public buildings of the old city of Tartous and the equipment locations: **a** the public multifunctional gallery (G), **b** the plan of the ground floor (right) and first-floor (left) of the Technical Office (T), **c** the plan of the ground (right) and first floor (left) of the Department of Antiquities (D). *Source* Author based on the schemes of Antiquities and Museums Department in Tatous

Fig. 5.8 Facades of the historic public buildings: **a** the eastern, western, and southern façades, respectively, of the Technical Office (T), **b** the northern and western façades of the Department of Antiquities (D), **c** Print shop (P)

type by covering the courtyard with a concrete ceiling. The second floor was built around the middle of the nineteenth century in a similar plan of the first floor at that time (with a central hall). By comparing the plans and facades of both selected office buildings and the typical building style throughout Syrian historical periods, it is clear that the building facades and designs have been transformed and influenced by the common building style in the simultaneous building periods. The print shop (P) is located on the ground floor of a residential building near the Knights' hall with sandstone walls, a western facade, and an open view to the old city square. All information about the materials and elements of the selected historic office buildings are shown in Table 5.1, and an overview of all the measured rooms is shown in Table 5.2.

5.1.3 Zoom in Approach and the Microclimate Survey

This survey used Zoom In approach to collect the subjective and objective data for the first three steps (Define–Measure–Analyze) of the Six Sigma circle (DMAIC/DMADV), and to provide the basic vision for further research and steps of design, control, and verification, as shown in Fig. 5.9. For the investigation and identification of the data needed to survey the selected structures of Tartous old city, two data collection methods were used; a questionnaire as a subjective measurement was used to obtain the occupants' responses to their environment and physical measurements were used to obtain quantitative data on the actual conditions in the buildings.

Table 5.1 Materials and elements of the selected historic office buildings

	Department of antiquities in Tartous (D)	Technical Office of the old city of Tartous (T)
Location and building orientation	In the waterfront of the old city of Tartous, they rise about 1 m above the sea level Both buildings are oriented to the west (to the sea)	
Shape and Floors	The selected buildings share the cross-vaulted ground floor. The ground floor is used as a public multifunctional gallery	
	It is situated on two floors. First floor (eight rooms and inner yard) contains two parts; the first part contains the main block of four rooms and the other part is with four rooms surround an inner yard. The second floor is built above the main block and contains eight rooms	A rectangular shape with central hall plan on the two floors. First floor contains five main rooms and the second floor contains five main rooms
Material and construction	The first floor rooms have two kinds of roofs. The rooms of the Ottoman parts have sandstone cross-vaulted roofs. The rooms of the other periods have concrete flat ceiling with exposed iron and wooden beams. The second-floor rooms just have a concrete flat ceiling. The roofs are carried on sandstone pillars for the Ottoman parts and concrete columns for the other parts of the later period	
Courtyard	It has an inner yard in a part of the first floor	No courtyard
Walls type	Sandstone walls of single layer	
External walls thickness	25–35 cm	25–30 cm
Internal walls thickness	25–20 cm	25–20 cm
Windows type	Two casements can wide open made of a single pane of glass and wooden frame with wooden shutters. Some of the windows have thin iron bars	
Windows shape	The external windows of the waterfront are simple rectangular openings and there are three arch-ending upper windows in the center of the second floor. The internal openings have a rectangular shape The other external façade openings are rectangular shape with wooden shutters	The external windows of the waterfront are simple rectangular openings with the arch-ending frames and there are three arch-ending upper windows in the center of the second floor. The internal openings have a rectangular shape. The external eastern façade openings are with hollow wooden shutters

(continued)

Table 5.1 (continued)

	Department of antiquities in Tartous (D)	Technical Office of the old city of Tartous (T)
Exterior doors	The main gate is wooden double panel door with external iron bars, and the upper opening of it is arched shape of single pane glass and iron bars	The main gate is wooden single panel door with upper open-arched opening
Interior doors	The internal doors are made from wood with different sizes	The internal doors of the first floor are made from wood with almost a same size. The internal doors of the second floor are the same-size wooden doors with upper colorful glass part
Insulations	The external skin of the façades is without insulation (natural sandstone)	
	The internal skin of walls and cross-vaulted roofs of the first floor are coated by a traditional white lime plaster. Whereas the internal skin of walls of the second floor are coated by the lime-concrete plaster, as well as its ceiling	The internal skin of walls is without insulation and just has the natural texture of sandstone The ceiling of the second floor is coated by the lime-concrete
Ancient systems and sun shading methods	No ancient systems. The sun shading methods are curtains and shutters. Balcony of the French period is removed from the center of the second-floor façade, because it was considered as transgression	
Energy upgrades	Most rooms have electric and Mazut heating devices and fans in summer. Air conditioners are used in three rooms	The rooms have electric fans and heating devices. Air conditioner is just used in the main room of the first floor

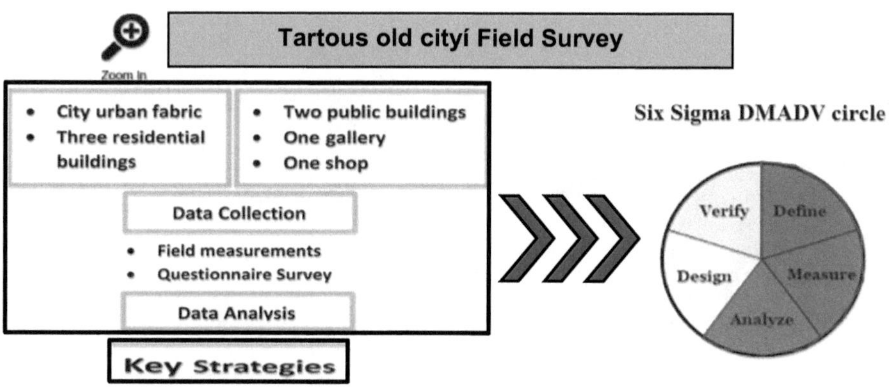

Fig. 5.9 Methodology structure

Table 5.2 Overview of all the measured rooms in the selected historic office buildings

Building tower	Floor	Room name	Room area (m^2)	Openings								Notes
				Internal façade openings				External façade openings				
				Internal orientation and ordinary window number	Upper window	door	openings to wall ratio (OWR) (%)	External orientation and window number	Upper window	door	openings to wall ratio (OWR) (%)	
The department of Antiquities in Tartous (D)	F1	Diwan Room (D-F1-DIR)	32.53	1	–	1	16	West(2)	–	–	17	The rooms (ER and TR) are open from two opposite sides, where their external facades are oriented to the west and the opposite sides are open to the inner yard from the east
		Director Room (D-F1-DR)	52	1	–	2	23	West(3)	–	–	13	
		Engineers Room (D-F1-ER)	39.43	East(2)	1	1	18	West(3)	–	–	14	
								South(2)	–	–	14	
		Technical Room (D-F1- TR)	32.05	East(1)	–	1	13	West(3)	–	–	15	
	F2	Buildings Room(D-F2-BR)	28.45	2	3	1	61	West(3)	3	–	46	
		Lobby Room (D-F2-LR)	50	3	3	6	13	North(2)	–	–	15	
The Technical Office of the old city of Tartous (T)	F1	Director Room (T-F1-DR)	37.65	2	–	1	14	West(3)	–	–	14	The central rooms (DocR and ADocR) are open from two opposite sides, where their external facades are oriented to the west and the opposite sides are oriented to the east. The worker room has a balcony from the eastern side
		Study Room (T-F1-SR)	36.53	2	3	2	20	West(3)	–	–	14	
		Documenting Room (T-F1-DocR)	81.27	–	6	6	8	West(3)	–	–	15	
								East(2)	2	–	17	
	F2	Worker Room (T-F1-ADocR)	69.17	–	–	8	16	West(3)	3	–	39	
								East(3)	3	–	45	

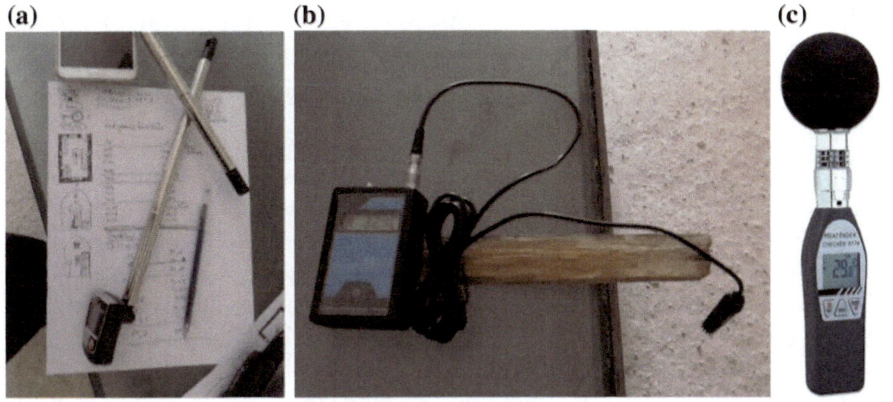

Fig. 5.10 Study cases' measurements equipment: **a** Testo 405-v1, **b** temperature and humidity data logger, **c** AZ8778 digital handheld black bulb thermometer

① Field Measurement (Objective Measurement)

To determine the thermal comfort level of the indoor environment for buildings, a survey study was carried out into two public buildings, three residential houses, and one shop. The buildings were monitored during August 2016. Each building was surveyed for seven days continuously. The measurements include the related thermal comfort parameters (relative humidity, air temperature, wind speed). The indoor and outdoor wind flow was measured three times in a day. Two personal parameters (clothing value and metabolic heat) were calculated and fixed according to ASHRAE Standard-55 and ISO 7730. The participants were surveyed in their structures, and the recording devices were placed at the height of the occupant at the seated level at 1.1 m. The devices were set to record the thermal parameters during the period of approximately 10 min. In general, the field measurement procedures in the old city of Tartous used the same procedures of Arwad island procedures but with more devices and in different observation time. The old city of Tartous measuring equipment included eight temperature and relative humidity data loggers, two Testo405 anemometers for measuring the indoor–outdoor airflow and temperature, and two AZ8778 digital handheld black bulb thermometers, as shown in Fig. 5.10.

(1) The Measurement of the Residential Historic buildings

The three residential buildings were monitored during August 19–25, 2016. For house one (H1-K), the data loggers were hanged in the living room (LR) of the first floor and bedroom (BR) in the second floor, and both rooms have south orientation. In the house two (H2-M), the instruments were hanged in the living room (LR), bedroom 1 (BR1) and bedroom 2 (BR2), all the rooms have south orientation. As well in the house three (H3-S), the instruments were hanged also in the living room (MR) north orientation and in the entrance room (SR) with the south orientation, refer Fig. 5.5. However, the number of dwelling monitored is small compared to a total number of building which are in the case study area, this is due to the shortage of time but the thermo-hygrometer devices were placed in different rooms' types and orientations.

(2) The Measurement of the Public Historic Buildings

The historical office buildings were monitored continuously for seven days from August 3–10, 2016. The public multifunctional gallery was also monitored for one week from August 10–17, 2016, and the print shop for one week from August 18–25, 2016. The data loggers were hung in four rooms in building (T): the central hall of the first floor (DocR), Director Room (DR), Study Room (SR), and the central hall of the second floor (ADocR). In addition, the equipment was hung in six rooms in building (D): the Diwan Room (DIR), Director Room (DR), Engineers Room (ER), and Technical Room (TR) on the first floor and the Buildings Room (BR) and Lobby Room (LR) on the second floor. Additionally, four points were chosen to fix the data loggers in the multifunctional gallery (G) and one in the print shop, refer Fig. 5.7.

② Questionnaire Survey (Subjective data)

The purpose of the thermal indoor environment survey is to evaluate the thermal conditions in the historic buildings in Tartous old city, through the information collected from the occupants of these buildings. The questionnaire is used as a tool of research providing a link between the objective of the subject and the respondent's information. The questionnaire designed based on ANSI/ASHRAE 55-2004 thermal comfort standard and other previous studies (Nematchoua et al. 2014; Buratti and Ricciardi 2009; Ricciardia and Buratti 2015; Ricciardia et al. 2016; ASHRAE, ANSI 2004). The questionnaire is divided into four parts and used Likert Scale of five- and seven-point scales and sensation voting scales, see Figs. 5.11 and 5.12 and Appendix C:

Part 1 (Basic and personal data and demography of respondent) consists of general information of the participant included the name, age, gender, years of

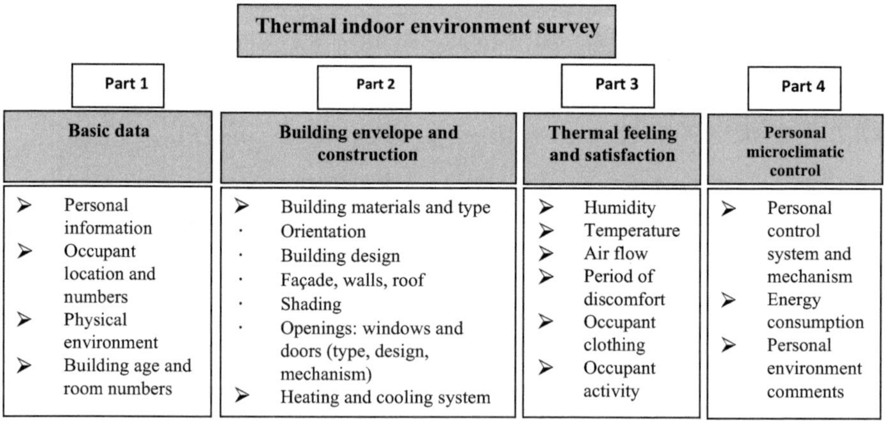

Fig. 5.11 Content of the thermal questionnaire

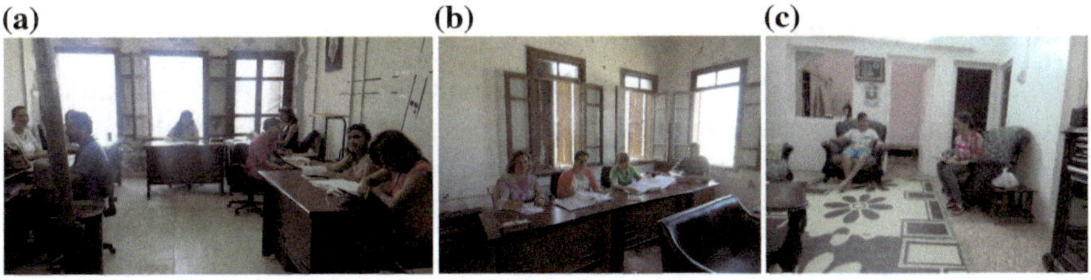

Fig. 5.12 Field measurement and the thermal questionnaire in the old city of Tartous: **a**, **b** The employees in the Antiquities and Museums Department in Tartous (D). **c** One of the houses' respondents

living in their current building, physical environment, and location. The main indoor thermal parameters were measured and registered during the survey.

Part 2 (Building envelope and construction) describes the characteristics, features, and components of the surveyed buildings; such as opening (door, windows types and materials), height of the building, walls, and roof materials. It also includes the use of different thermal environment control means (windows, local fan/AC/shading devices).

Part 3 (Thermal feeling and satisfaction) records the respondent's overall satisfaction and votes in term of the indoor thermal variables (air temperature, relative humidity, and air velocity), and documents the recent activities and clothing.

Part 4 (Personal microclimatic control) includes personal control system and mechanism, energy consumption, and two open questions. The energy consumption questions were excluded due to the war circumstances during the survey period.

The sample size of the questionnaire in the selected public buildings was 70 respondents (12 questionnaires to the Technical Office of the old city of Tartous (T), and 58 questionnaires to the department of Antiquities in Tartous (D), which is 80% of the total number of the employees in both selected office buildings.). Additionally, the questionnaires were randomly distributed in 58 dwellings covering all Tartous old city area (the waterfront, Alsahla, Donjon, the square, the external and internal walls houses, Al Khandak, the church Arafat basement neighborhood, the Knights Hall houses) to provide more generalizable results for the statistical point of view. The target sample size for the questionnaire was 191 subjects which is 13.3% of the total population of this case study area. All the participants are those residences who lived in the old city of Tartous. All respondents were given 15 min to answer the questionnaire and retain it. Two observers were played part in this operation. The questionnaire was conducted over a series of four weeks period in the summer season of Tartous of August 2016. The questionnaire took place between 10 am and 3 pm.

③ **Data Compilation and Analysis**

All the physical measurements of the environment and data from the questionnaire surveys were analyzed by placing information into sheets, creating tables, pie chart, and excel spreadsheets. The software SPSS was also used to analyze the thermal satisfaction of the old city occupants and the thermal conditions in their structures. Analysis was presented into three parts; results of the questionnaire, results of the field measurements, followed by a comparison with the previous studies and with the standards of ASHRAE and ISO. Then, getting into the treatment and key strategies.

5.2 Microclimate Field Survey of the Old City of Tartous' Dwellings

5.2.1 Questionnaire Survey Results

① **Demographic and Basic Information**

The population of Tartous old city is about 3500 people, but the real population is 1438 inhabitants. The difference is due to travel abroad and migration of the majority and housing outside the old city. It has high unemployment where the employees in the city are 28.5, 11.9% non-employed, and 71.5% non-productive consumers, and a low level of education in the old city with 12.3% of illiterate and 68.6% of learners with basic education (Bashour 2011). All of the respondents of our survey have not a professional activity related to cultural heritage. The dominate gender samples were females 102 (53.4%), and 89 (46.6%) were males. More than half of the respondents were the young category and this gives a further opportunity to work on raising awareness and changing their adjustment behavior.

The respondents of the sample size have a low level of education with 10.47% of illiterates, 67.55% of learners with a basic education, 15.7% of secondary education, and 6.28% of high education. The sample size has high unemployment where the non-employed and housewives in the city are 47.64, 17.28% of non-productive consumers (students), 8.37% of them have sea and maritime works, 8.38% of employees and retired, and 18.33% with other works. The total number of occupants in one house is as follows 29.31% of families with less than three persons, 55.18% of families with four to six persons and 15.51% of families with more than seven persons. Most of the families live in their house for more than 30 years, as shown Fig. 5.13.

② **Building Envelope Evaluation**

(1) **Building Envelope and Materials**

The old city of Tartous possesses three types of houses, unequally represented by the house with an inner courtyard (a house belongs to the ottoman and late Ottoman periods and before the middle of the nineteenth century), the house with a central hall (a house is built in the French mandate period and after the middle of the nineteenth

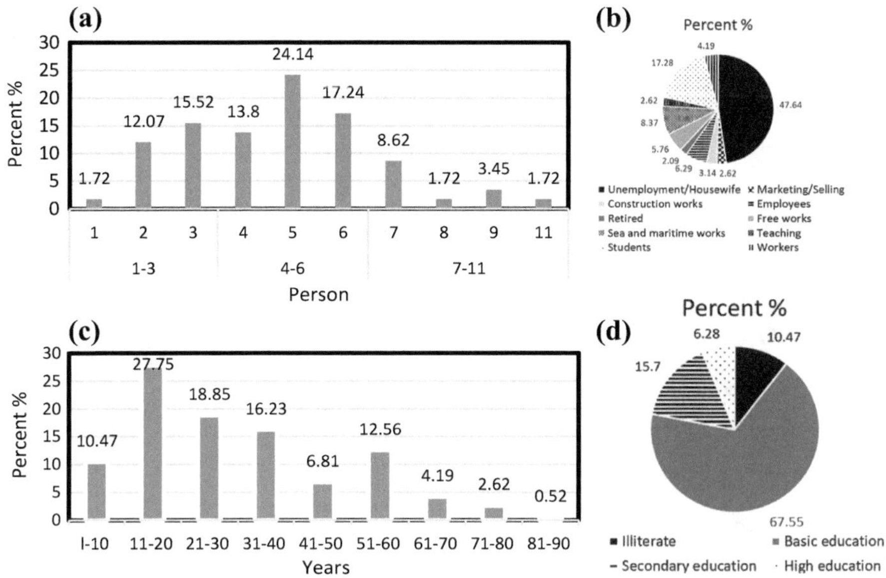

Fig. 5.13 **a** Total number of occupants in the house, **b** respondents' occupation, **c** the residence duration of the respondents, **d** respondents' education

century), and the house without inner courtyard or central hall. In our survey, the main architectural features of the selected houses belong to the Late Ottoman and the French mandate periods (between the nineteenth century and the first half of the twentieth century). The houses belong to the Ottoman period are about 37.93, 39.66% of the surveyed houses are from the French mandate, 17.24% for more than 25 to less than 70 years old and 5.17% less than 25 years old. Most of the houses have more than three rooms. A total 44.8% of the select houses located in the first floor, 25.9% in the ground floor, 15.5% in the second floor, 5.2% in the third floor and 8.6% distributed in two or three floors, as shown in Fig. 5.14. From the review of the results of the survey, 13.8% of the houses have been built by using the industrial stone and concrete, while 68.52% of the respondents' results show that their buildings have been built by using sandstone, and 20.68% of their houses have been mixed sandstone with other materials. Most of the houses are without courtyards (81%) and a total 6.9% they roofed over the courtyard and made it an extra room, 6.9% with an inner courtyard, and 5.2% of houses have outside yard. Most of the houses have more than two external facades, 34.48% of them just have one external facade, 1.72% is separated block and open from all the facades, and one block is without external facades. The external walls' type of the surveyed houses is 63.8% of sandstone walls of one-layer with a thickness of 20–40 cm, 13.8% of the massive sandstone walls of double-layer wall with a thickness of 70–200 cm, and 3.45% of the massive sandstone walls of single and double-layer walls. A small percent 13.75% of the surveyed houses have just artificial stone walls (concrete blocks walls), and 5.2% of

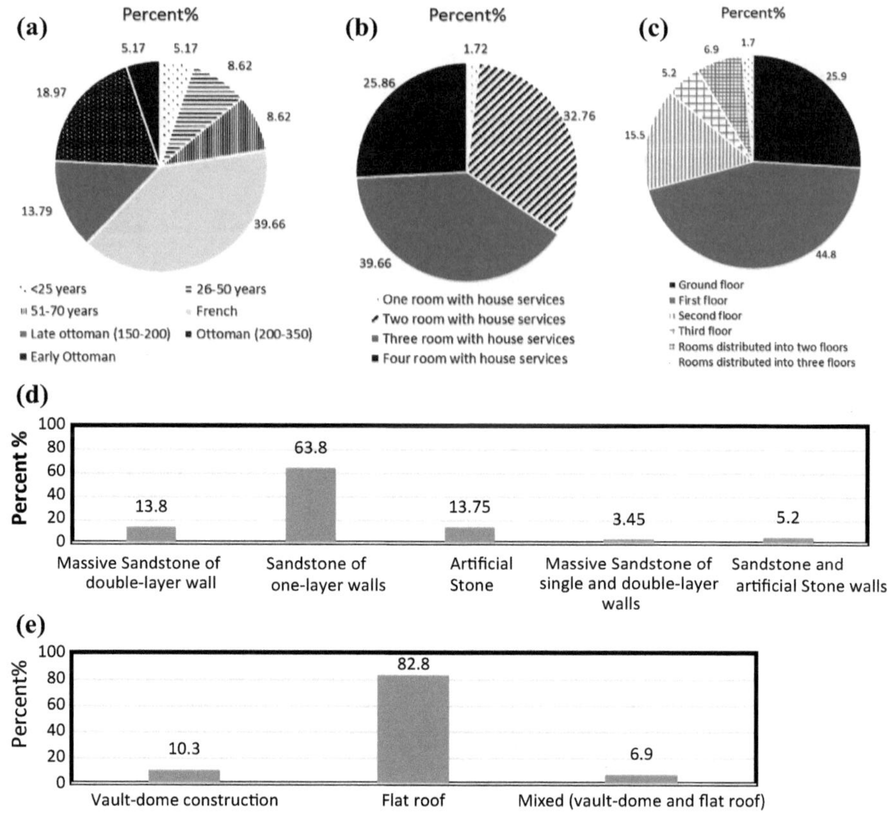

Fig. 5.14 Basic data of the surveyed residential buildings: **a** the age, **b** the number of rooms in the house, **c** house floor, **d** the type of the external walls, **e** the roof type

houses have mixed sandstone and artificial stone walls. A total 50% of the external walls insulation are concrete plastered, and 5.2% of concrete plastered insulation painted by white color, and just one surveyed house has external walls painted by yellow color, while 44.8% of external sandstone walls are without insulation.

Nearly half (51.7%) of the internal walls thickness varies between 10 and 20 cm, the other half thickness (44.9%) varies between 25 and 30 cm, and just 3.4% of the internal walls have a thickness varies between 30 and 50 cm. A total 77.6% of the internal walls covered by concrete plastered, and 19% of the internal walls covered by concrete plastered and painted by white color, and 3.4% of the internal walls stayed without insulation. The roof type mostly is a flat roof, 10.3% is still using the vault-dome construction which is the traditional mode of the ceiling that helps to minimize the amount of heat gain to penetrate inside through the roof, and 6.9% contained a vault-dome and flat roof together. Nearly two-thirds of the houses' roofs are made of concrete, 10.35% of the houses' roofs are made of concrete and wood, and 10.35% of the houses' roofs are made of sandstone. 6.9% of them contained a sandstone and concrete. A minor percentage of the houses' roofs have aluminum plates, gypsum, and hardboard roof. Most houses use electric heating devices in winter and a total of 37.9% of the city respondents use local organic material for heating such as wood and timz which is the dried molds of the collected residues of the olive seeds and oil, and approximately 15.5% of them use gas for heating. Air conditioners are rarely used in the living room in less than 30% of the surveyed houses in summer. Most of the historical buildings have natural ventilation. Electric fans and manual devices are the most used tools in the hot periods in the houses of the old city of Tartous.

(2) **Opening and Sun Shading Device**

A total 24.14% of occupants' dwellings have the openings between (1–3), while 65.53% having openings between (4–6) and remaining 10.33% having (7–11) openings in their buildings. The windows' materials are glass with wooden or aluminum frame. The occupants were asked to identify the type of windows; a total 60.35% use casement windows, while 25.85% use aluminum sliding windows and 12.1% use both latter windows together (casement and aluminum sliding window) and one house does not have windows. All of these windows do not help to prevent sun ray's penetration inside the building, and the aluminum sliding window prevents half of ventilation, as shown in Fig. 5.15. A range of sun shading device was listed on the survey for respondents to indicate which shading method they use and to specify a reason of using that shading. All the surveyed houses are without balcony. A total 70.7% of the houses' windows are without shutters and 6.9% of windows just have iron bars, while the other houses have wooden casement shutters with 12.1 and 5.2% have rolling shutter 3.4% have wooden shutters our traditional mashrabiya with iron bars. Due to the lake of shutters, most residents use curtains (82.8%), and 5.2% of houses keep the shutters closed to provide the privacy, 13.8% of them they do not have any sun shading method and no one responds on other shading devices apart from these methods. As the curtains were the dominate choice for occupants, the reasons given were as following:

(1) Curtains are the most straightforward and affordable way to provide the privacy which the respondents were accustomed to.
(2) Curtains have its positive function as the decorative elements in the room, and they keep furniture from direct sunlight and provide sun radiation protection.
(3) Additionally, there was no other option because the building code of the old city avoids putting overhangs on the top of the windows or changing the original shape of the traditional windows.

③ **Indoor Environment Parameters Assessment**

(1) **Indoor Air Temperature and Relative Humidity**

The overall indoor air temperature taken in each dwelling during the questionnaire survey ranged from a low of 28.7 °C to a high of 33.6 °C and the indoor relative humidity ranged from 54.4 to 75%. And in the meanwhile, the inside airspeed ranged from 0.03 to 0.18 m/s, and the radiant temperature ranged from 30.67 to 35.68 °C, as

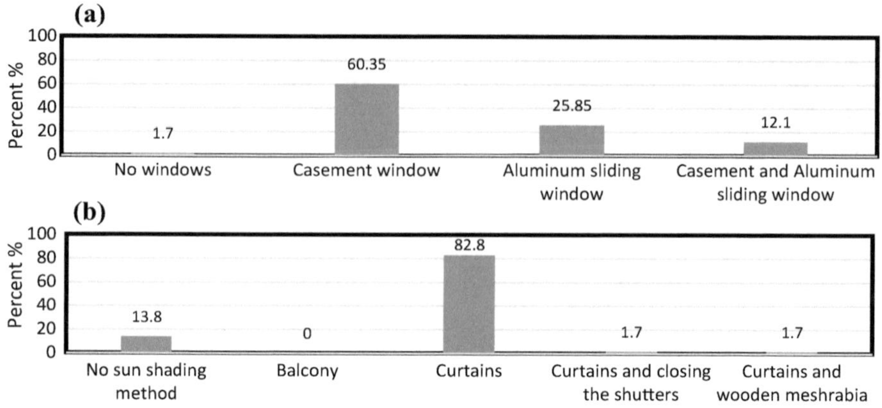

Fig. 5.15 Openings and shading methods of the surveyed residential buildings: **a** windows style, **b** the methods of the sun shading

Table 5.3 Indoor air temperature, relative humidity, and wind speed of the dwellings during the questionnaire survey

Occupant location in area		Inside air temperature (°C)	Inside airspeed (m/s)	Humidity (%)	Radiant temperature (°C)
Living room (LR)	AV	31.175	0.12	65.42	33.14
	MAX	33.600	0.18	75.00	35.68
	MIN	28.700	0.03	54.40	30.67

shown in Table 5.3. From these data, the operative temperature (OP) was calculated, and with the mean radiant temperature, air temperature and airflow that would produce the acceptable temperature (ISO 1998). The resultant of operative temperature averaged 32.11 °C for the entire sample, and according to ASHRAE Standard-55 (2004), the operative temperature average based on air and the mean radiant temperature was 32.16 °C (ISO 1998). It is obvious that OP obtained from the questionnaire survey was outside the limit of ASHRAE Standard-55 (2004), where the temperature ought to be between 23 and 26 °C and relative humidity should be between 50 and 60% (ASHRAE, ANSI 2004).

The occupants were asked how they feel in terms of humidity with a seven-point sensation scale, namely "Humidity Predicted Vote (HPV)", similar to ASHRAE Thermal Sensation Scale (2004) that specified the acceptable relative humidity should have 80% of occupants vote for the central three categories (−1, 0, +1). The results showed that only 27.6% out of 191 respondents (58 families) voted within the central three categories showing that the occupants were not in thermal acceptable condition within their dwellings according to ASHRAE standard (ASHRAE, ANSI 2004), Fig. 5.16. This means that about two-thirds of the occupants' overall vote was not satisfied/uncomfortable with the indoor relative humidity. A 5.2% of occupants feel

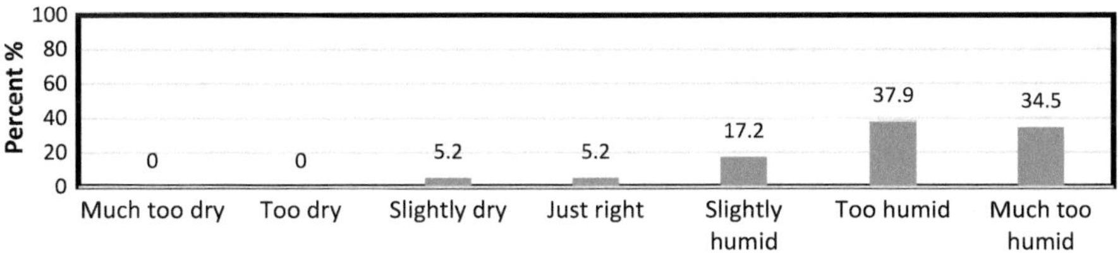

Fig. 5.16 Indoor Humidity Perception Vote (HPV) result in the residential buildings

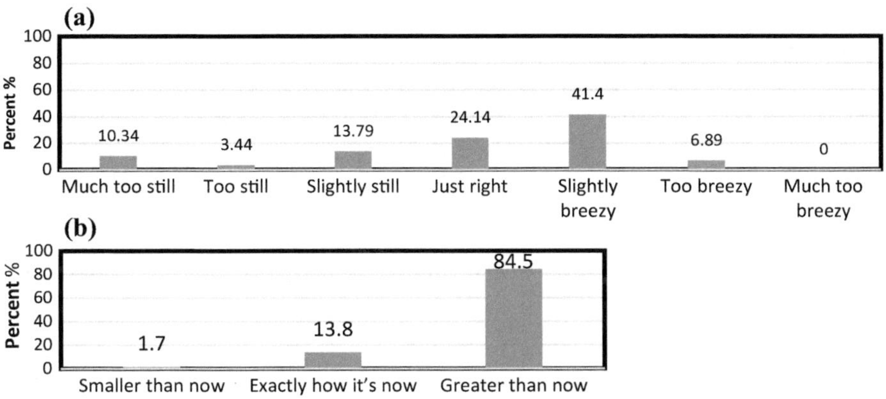

Fig. 5.17 Air velocity vote results in the residential buildings: **a** indoor Draft Perception Vote (DPV) results, **b** the respondents' preference of the air velocity

slightly dry another 37.9% feel too humid and remaining 34.5% feel much too humid. This conclusion shows that residents were too sensitive to humidity variations which show uncomfortable for the occupants.

(2) **Air Velocity**

Figure 5.17 displays the Draft Perception Vote (DPV) results, a vote −3 indicated that the air is perceived stagnant or motionless, a zero vote means that respondents feel the velocity just right. A total 13.79% of the subjects felt the air too steady (−3 to −2) while 79.33% respondents found the air velocity was proper (−1, 0, +1) and only 6.89% found the airflow ought to be less. These results show the acceptance of the air movement in their buildings. By asking the respondents directly about the preference of the air velocity, most of them (84.5%) preferred to increase the air velocity because of the high relative humidity.

(3) **Clothing Insulation**

The field survey reviewed how occupants dress indoors to assist in determining how they adjust to the temperature levels, for reason of becoming more comfortable. Male occupants wore three types of clothing and the outfits as calculated according to ASHRAE Standard 55-2004 sheet. The results show that their clothing value ranged

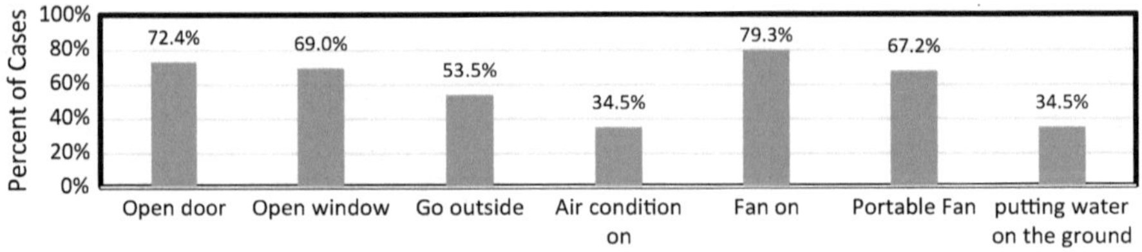

Fig. 5.18 Occupants adjustment/control choice

between 0.33 and 0.5 Clo units (short-sleeve shirt/blouse with or without underwear T-shirt, light trousers or walking shorts, and sandals or shoes). The reason provided for choosing these outfits is to feel cooler and more comfortable and made them feel cooler. Most female occupants were housewives or stayed at home and wore comfort house cloth (long-sleeve thin shirt dress for old women and short-sleeve thin pajamas for young women), the values associated with these kinds of clothing are between 0.39 and 0.48 Clo units as calculated according to ASHRAE Standard 55-2004. The clothing insulation value was in a wide range between 0.5 Clo for male and 0.48 Clo for female. The low value is near the value of 0.5 Clo suggested for the summer season in the ASHRAE Standard 55-2004.

(4) **Result on Personal Adjustments**

The respondents to the survey were asked the method used to adjust or control their buildings. Figure 5.18 shows the multiple response questions about the personal adjustments. A total 34.4% of respondents preferred to be cooled naturally, with 27.7% preferred mechanically means (AC/Fan). Another 24.8% controlled themselves by different methods (taking shower and reducing clothing, portable fan, or putting water on the ground), and 13.1% go outside their houses. The preference of the natural ventilation was because of several reasons; firstly, because the natural ventilation provides a comfortable microclimate in the ventilated space and brings fresher and cleaner air into the living environment. Secondly, it is healthy and comfortable for the human being more than applying the mechanical ventilation system which could cause them sickness. Thirdly, the natural ventilation is more affordable in comparison with the electrical bills of the air conditioning which highly increased in Tartous nowadays. In the other hand, the preference for mechanical means was because the respondents think that the mechanical ventilation provides more control when there is no control over the source or amount of air that comes into the house. It offers proper coolness at any time along with the appropriate locations.

④ **Thermal Response Vote Evaluation**

(1) **Thermal Sensation Votes (TSV)**

The results of subjective responses to temperature (thermal sensation) are shown in Fig. 5.19. The results clarify that the plurality of the residents voted between "**warm and hot sensation**". A total 10.34% of the respondents found their environment

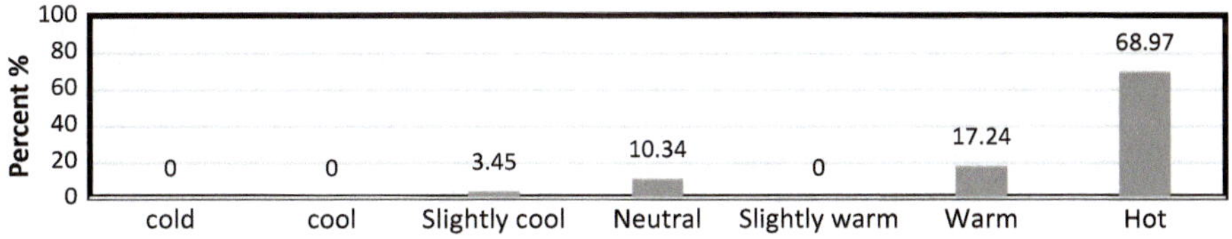

Fig. 5.19 Thermal Sensation Vote (TSV) results

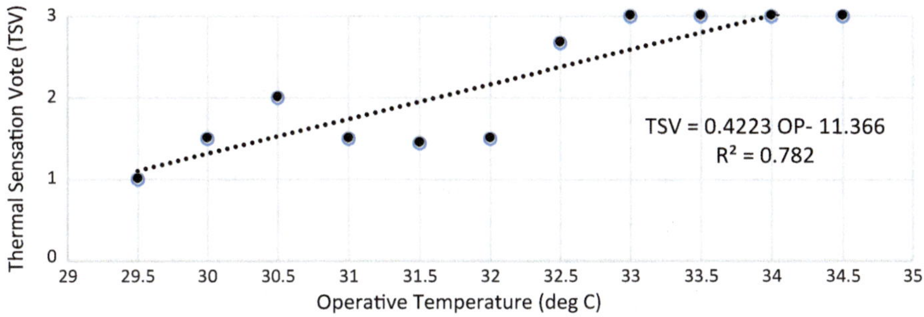

Fig. 5.20 Linear regression of thermal sensation votes versus operative temperature in the surveyed houses old city of Tartous

neither cool nor warm in the neutral ground, 17.24% of them being warm preferring cooler conditions, 68.97% of them experience the hot intolerable condition, and 3.45% of them found their environment to be slightly cool. In this study, 13.79% out of 191 respondents (58 families) voted within the central three categories, 1(slightly cool), 0(neutral), and 1(slightly warm), showing that occupants have not thermal acceptable conditions within their buildings (ANSI/ASHRAE 1992).

The relationship between thermal sensation and operative temperature forms the basis of research into thermal neutral temperature, and an organized method to predict the subjective comfort which results from a given temperature is regression analysis. Simple linear regression was performed between thermal sensation and operative temperature to specify the strength of the relationship between them in the residential buildings. The data have been binned in the interval of 0.5 °C and plotted as shown in Fig. 5.20. A regression line was then fitted to predict the thermal sensation of the entire points of the monitored temperature. The fitted regression equation for subjects' sensation versus operative temperature was:

$$\mathrm{TSV}(R) = 0.4223\,\mathrm{OP} - 11.366 \quad R^2 = 0.782 \tag{5.1}$$

**Correlation is significant at the 0.01 level (two-tailed) for 11 groups after using BIN method.

The coefficient of determination (R^2) between the thermal sensation votes and the operative temperature is 0.782. Therefore, about 78.2% of the variation in the thermal sensation vote data is explained by temperature. The value obtained of R^2 indicates the significance between the observed/predicted thermal sensation and the indoor operative temperature; this value shows a strong relationship between dependent and independent variables. The thermal sensation votes correlated strongly with the operative temperature shows a good agreement between TSV and OP. The slope of the regression line represents the sensitivity of the subjects with respect to the operative temperature. Therefore, the small value of the regression line slope 0.42 °C demonstrates that occupants of the old city of Tartous are tolerant of a wide range of temperature. The neutrality condition is derived by solving the regression equation for zero (neutral) yields an estimated neutrality of 26.42 °C. A neutral temperature of 26.42 °C and a comfort band of (24.05 and 28.78 °C) coinciding with (-1 and $+1$) sensation vote have been obtained using this regression equation. The finding found to be more tolerant for hot conditions than the range of temperatures (23–26 °C) specified by ASHRAE Standard -55.

To investigate the obtained summer comfort band of the old city of Tartous, the neutral temperature equations of de Dear et al.', Auliciemes, and Humphreys were utilized. Based on the climate data of the coldest and warmest month (January and August) of our case study site Tartous, the predicted the neutral temperature was found as shown in Table 5.4, refer Eqs. (2.1)–(2.4). Auliciemes Equation overestimates the thermal neutrality temperatures for the winter, as well as the above methods, was just trustworthy in the hot period more than the cold period and can be used to verify the obtained summer comfort band. The results of our survey obviously comply with the results of the summer comfort band of above methods. The thermal comfort zones depend on the regional climate. There is a limited amount of literature on human comfort and on historic or office buildings in moderate or eastern Mediterranean climates. To verify the resulted comfort band of our survey, a comparison with the previous studies in a similar climate were done. The obtained neutral temperature of this survey has similar value of the summer neutral temperature (26 °C) of a field study conducted in Haifa, Palestine for non-air-conditioned dwellings (Becker and Paciuk 2009).

(2) **Thermal Preferences**

The results of the subjective thermal preference among the occupants are shown in Fig. 5.21. It is obviously seen that a high percentage of respondents 86.2% preferred to be a little bit cooler in their buildings, and a small percentage of them 8.6% want too cool indoor temperature. A 5.2% of the respondents indicated that changes in the thermal environment are not required, whereas no one stated that they prefer to be warmer as Tartous has a humid summer climate, so definitely, nobody would want to be warmer on summer period. The preferable condition should be a qualitative rather a direct factor to be used for the thermal design or structure assessment. The strategies considered the preferred condition rather than the neutral one can ignore the positive impact of the thermal adaptive opportunities. Considering the neutral instead of the preferred temperature provides an extension of the comfort zone and reducing the significant energy and environmental costs.

Table 5.4 Tartous comfort band, according to de Dear et al.', Auliciemes, and Humphreys natural temperature methods

Comfort limits (°C)	De Dear et al' equation		Humphreys' equation		Auliciemes' equation		This survey
	Tn = 17.8 + 0.31 Tm		Tn = 11.9 + 0.534 Tm		Tn = 17.6 +0.314 Tm		TSV(R) = 0.4223 OP-11.366
	Winter	Summer	Winter	Summer	Winter	Summer	Summer
Upper temp (TU) = Tn + 2.5 °C	24	28.5	20.8	28.6	23.9	28.4	–
Neutrality tem (Tn) =	21.5	26	18.3	26.1	21.4	25.9	26.42
Lower tem (TL) = Tn_2.5 °C	19	23.5	15.8	23.6	18.9	23.4	–
Comfort zone Tn ± 2.5 °C	19–24	23.5–28.5	15.8–20.8	23.6–28.6	18.9–23.9	23.4–28.4	24.05–28.78

Where Tartous winter Tm = 12 °C and Tartous summer Tm = 26.5 °C

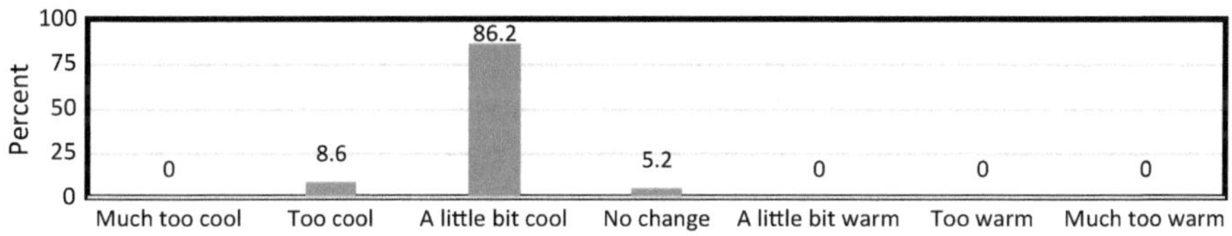

Fig. 5.21 Thermal preference vote results

(3) Thermal Acceptability

The surveyed participants were asked the rate of the overall acceptability. Half of the participants judged their thermal sensation are not acceptable, 65.52% of them explained the reason for the dissatisfaction that the temperature always too hot, and 34.48% of them find it often too hot. The other half of the participants accepted the overall temperature of their building, and 72.4% of them had a reservation on the temperature that is occasionally too hot. It seems that for those who desire to be cooler are the ones who vote on the unacceptability of the air temperature of their dwelling and the subjects responding in the neutral zone, thus having thermal comfort are the ones who are satisfied with the temperature of their buildings. These results show that thermal acceptability is related to people's thermal preference. They also demonstrate that those respondents, who preferred "no change" to their environment,

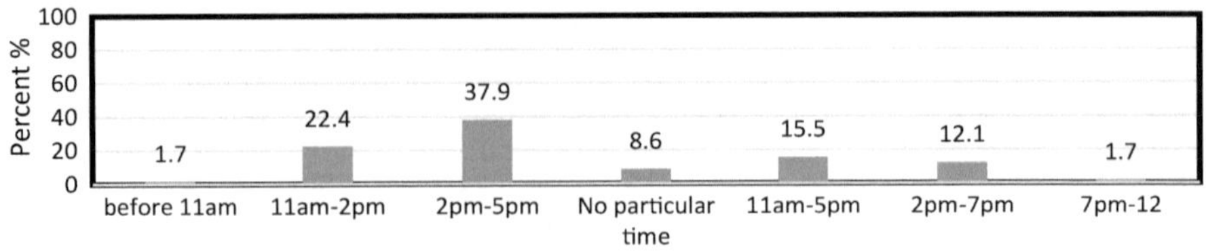

Fig. 5.22 Discomfort time in the surveyed houses

were not always having "neutral" thermal sensation. According to the respondent's votes, the discomfort time in the surveyed houses was mid-day and afternoon from 11 am to 5 pm, as shown in Fig. 5.22.

⑤ The Results of the Open Questions

Thermal comfort is one essential issue from several other important needs of the local people in their properties. The term "thermal comfortable condition" meant to the residents "feeling comfortable in the house to live and do house works and daily works effectively", "to be comfortable", "have cool spaces and suitable temperature and airflow", and "moderate and acceptable internal atmosphere". For some respondents, it meant "changing room temperature to be comforted by heating in winter and cooling in summer". According to elder's interviewers, the residents are not satisfied with the thermal condition in their houses during the last 30 years; they said that before they were comfortable with the indoor environment but now things are completely changed. They were nevertheless familiar with the situation and knew how to cope and adjust for the climate variation and continue satisfied with their indoor thermal environment". That shows that the residential historic buildings are already affected by climate change issues. A total 70% of residents mentioned the need of using heating and cooling systems, and wanted to use cooling devices rather than the natural ventilation for their buildings to ensure thermal comfort, which will increase the energy consumption and may affect their traditional structures.

In summary, the occupants commented in the open comments about their houses environment that they suffer from the high temperature and humidity during summer. They need roof insulation, shutters and shading methods, the maintenance of the walls and mortar which cause the sand and lime dust. They need besides to improve the environment and the code of old city buildings. Two answers asked the help of changing their house and repair the dilapidated ceiling. The current fenestration does not provide privacy and ventilation. Therefore, the occupants asked the help of the Technical Office of Tartous municipality to modify the old city code to find or design the fenestration and shading methods to meet the needs of the population and at the same time to beautify the city. The code of the old city of Tartous forced the removal of additional balconies of the residential buildings, then restoring the building to the historical situation, regardless of the different usages of the building in the past as a defensive use and as a dwelling now in line with the need for the residents, as shown in Fig. 5.23. Therefore, the residents requested to allow the rebuilding of the

Fig. 5.23 Current fenestration and the removal of additional balconies of the residential buildings because of the policy of the old city of Tartous

removed balconies, in particular, those found at the end of the last century which formed a kind of historical transformation of these buildings according to the needs of their inhabitants.

5.2.2 Field Measurement Results

Physical quantities equipment were an objective method in collecting data. In the field survey, three historical residential buildings, namely (H1), (H2), and (H3) were monitored. These environmental field measurements were done in seven consecutive days during 19–25 August, 2016. The survey was accompanied by measurements of the relevant thermal comfort parameters (relative humidity, air temperature, wind speed). The measurement results of the indoor air temperature, indoor relative humidity, daily average of the indoor and outdoor wind speed, daily average indoor and outdoor air temperature are shown in Fig. 5.24 (Table 5.5).

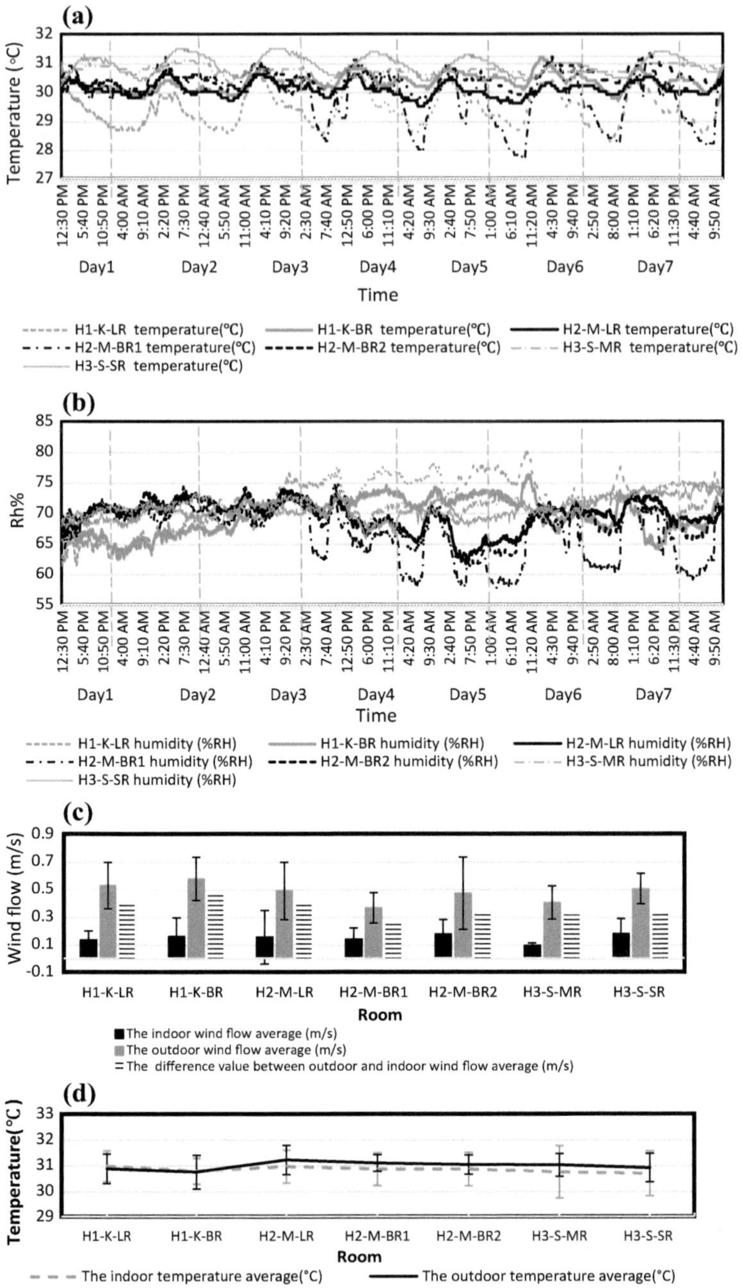

Note: vertical error bar refers to the standard deviation

Fig. 5.24 Measurement results of the three selected houses' rooms: **a** indoor air temperatures, **b** relative humidity, **c** August daily average of indoor–outdoor wind flow, **d** August daily average of indoor–outdoor air temperature

Table 5.5 Maximum and minimum indoor air temperature, humidity, and wind flow in each measured house

House	H1-K		H2-M			H3-S	
Room name	LR	BR	LR	BR1	BR2	MR	SR
Temp max	31	31.3	30.7	31.2	31.4	31.1	31.5
Temp min	28.2	29.7	29.5	27.7	29.7	30.1	30.2
Rh max (%)	80.7	76.3	73.8	74.8	74.8	75.3	75.2
Rh min (%)	61.5	63.2	61.6	57.6	61.5	66.2	65
Wind flow max	0.18	0.34	0.66	0.29	0.32	0.11	0.29
Wind flow min	0.09	0.07	0.04	0.08	0.1	0.05	0.07

The indoor air temperature in the occupied zone in the house one (H1) was varied between 28.2 and 31.3 °C with an average value of 29.75 °C and the indoor relative humidity between 61.6 and 80.7% with the peak value of 71.15%. The indoor air temperature as measured during the field survey in the house two (H2) was varied between 27.7 and 31.4 °C with a mean value of 29.55 °C. The range of indoor relative humidity in the house two was within 57.6 and 74.8% with the mean value of 66.2%. As well as the indoor air temperature as measured in the house three (H3) was varied between 30.1 and 31.5 °C with a mean value of 30.8 °C. The range of indoor relative humidity in the house three was within 66.2 and 75.2% with the mean value of 70.7%. A noticeable difference in temperature of approximately 2.8 °C between night and daytime can be seen in the living room of the first house, and in the second bedroom of the second house of approximately 3.5 °C between night and daytime. The second house living and first bedroom had a slight difference in indoor air temperature and relative humidity values between the nighttime and daytime. In the other hand, the third house' rooms have a relatively stable indoor air temperature and relative humidity values between the nighttime and daytime more than the other houses.

Considering the airflow in these dwellings, the first house rooms (0.13–0.15 m/s), and the second house rooms (0.13–0.17 m/s) had relatively good ventilation due to their orientation to the prevailing west southern wind in summer. The third house has a wind flow (0.09–0.17 m/s) because of the opposite direction of the house to the prevailing wind direction. The indoor air temperature in the measured rooms in the first house was more or approximately equal to the outdoor air temperature. The indoor air temperature in the second and third house rooms was less than the outdoor air temperature with a slight value of 0.2 °C. The temperature recommended by the ASHRAE standard should be between 23 and 26 °C and relative humidity ought to be between 50 and 60%. From the results obtained in all the houses, it can be observed that most of the temperature and relative humidity values transcended the standard for sedentary activity during the summer period.

① Variation in the Air Temperature and Humidity over the Measurement Intervals

The averages of the indoor air temperature and humidity in each day periods (intervals of times) of the all measured rooms for the historical surveyed houses (H1-K), (H2-M), and (H3-S) are shown in Fig. 5.25. The indoor temperature average of the bedroom of the first house (H1) rose steadily from 30.05 °C in the morning and reached a peak of about 30.58 °C between the hours of (11–7 pm), and the same happened to its relative humidity that ranged steadily between 68.8 and 70%. The indoor air temperature average of the living room rose from 28.9 °C in the morning and reached a peak of more than 30 °C between the hours of (11–7 pm), then the air temperature slowed down to 28.82 °C after 7 pm. The average of the relative humidity decreased during the noon and afternoon and increased after 7 pm to reach the peak of about 73.6% in the early morning. Both the bed and living rooms of the first house (H1) raised the air temperature in the noon and afternoon time with increasing the outdoor solar radiation, as well the bedroom showed rising of air temperature more than living room during the day intervals. The second house (H2) also showed increasing in the air temperature and decreasing in the relative humidity in the noon and afternoon time in proportion to the outdoor solar radiation. The bedrooms showed increasing the air temperature more than the living room from 11 am to 11 pm. The averages of the air temperature and the relative humidity in the second house (H2) showed that the indoor space at night is better than it during the daytime, especially in the first bedroom. In general, the third house (H3) had a relatively steady indoor air temperature and relative humidity during the day intervals compared with the other houses. It had the highest air temperature averages during the measurement period compared with the other houses. It reached 31.29 °C between 3 and 11 pm. Furthermore, the relative humidity increased at night time in the measured rooms.

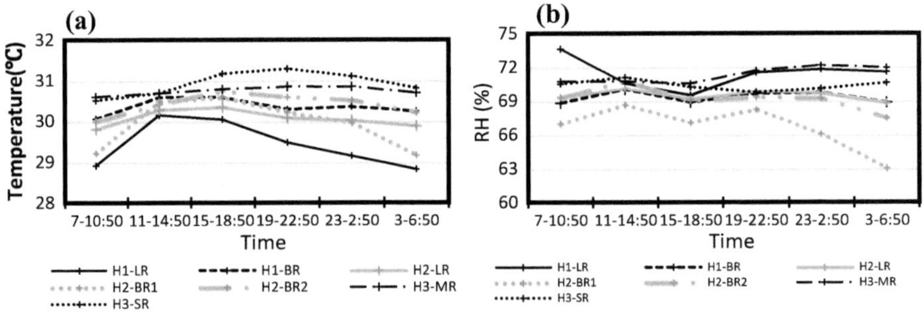

Fig. 5.25 Measurement data during the daytime intervals of the selected three houses: **a** the air temperature, **b** the relative humidity

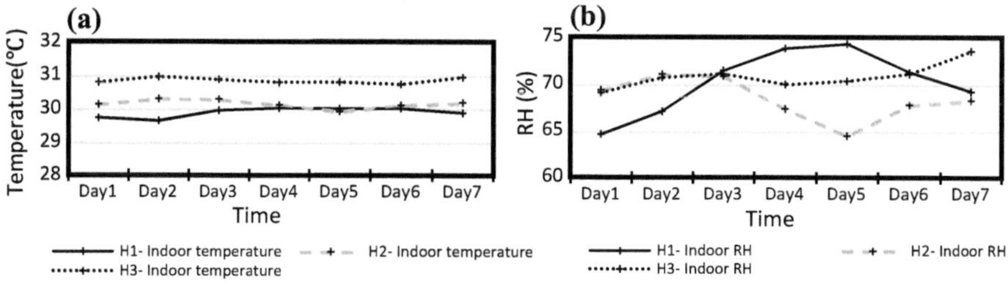

Fig. 5.26 Variation of indoor average air temperature and relative humidity of the (H1), (H2), and (H3): **a** indoor air temperature, **b** indoor relative humidity

② Variation in Air Temperature and Relative Humidity of the Three Houses (H1), (H2), and (H3) for Each Measured Day

Figure 5.26 shows the average indoor values of the houses for each measured day. The air temperature converged in the three houses by a slight difference of about 1.5 °C. The relative humidity is directly proportioned to the air temperature. The behavior of the relative humidity is not constant, it was less distress from day one to day two, and then it converged on the third day in the three houses. In day four and day five, it accelerated obviously compared with the related indoor air temperature values, then it back to decelerate for day 6 and 7 in the first and second houses. The occupants of the house (H1) are distress due to high humidity compared to the other houses, as it is shown on the graph for the daytime intervals and the graph for the third, fourth, fifth, and sixth day. The high humidity levels lead to discomfort due to the ineffectual evaporative cooling of the skin. The relative humidity in the second house (H2) showed better indoor conditions than the other houses. The third house (H3) had a relatively steady indoor air temperature and relative humidity and had the highest air temperature during the measurement period compared with the other houses.

③ Thermal Comfort Indices

According to ASHRAE55-2004, ISO7730, and the Chines standards) GB/T 18049–2000(and (GB 50736-2012), PMV should not exceed [−1, +1] range value, as well as the PPD maximum value, should not exceed 27%, refer. Tables 2.6 and 2.7. The averages of the measured indoor parameters (air temperature, relative humidity, air velocity) of our field measurement results in each studied room were used to calculate both of PMV and PPD. Considering dry bulb temperature equal the average of the measured indoor air temperature, the black globe temperature and the mean radiant temperature were calculated based on the equations of Table 5.6. Then, the personal parameter values (0.5 Clo) for clothing and (1 met) for the metabolic rate according to ASHRAE Standard-55 and ISO 7730, then CBE thermal comfort tool (Tyler et al. 2017) was used in calculating the PMV, PPD and finding the thermal sensation category for each room. To provide the acceptable and passive

Table 5.6 Results of calculating the thermal comfort indices (PMV and PPD), PMV modifications for the selected naturally ventilated historic residential buildings (H1, H2, H3)

Room name	Indoor air temperature (°C)	Indoor relative humidity (%)	Indoor air velocity (m/s)	TBS = Dry bulb temperature (°C)	TGN = Black globe temperature (°C)	Mean radiant temperature (°C)	Operative temperature (°C)	Clothing (Clo)	Metabolic Rate (met)	PMV	PPD (%)	Sensation	aPMV	PMVe
H1-K-LR	29.43	71.50	0.13	29.43	30.88	31.32	30.38	0.5	1	1.84	69	Warm	1.33	1.29
H1-K-BR	30.36	69.56	0.15	30.36	31.83	32.39	31.38	0.5	1	2.14	83	Warm	1.48	1.50
H2-M-LR	30.08	69.50	0.15	30.08	31.55	32.08	31.08	0.5	1	2.02	78	Warm	1.42	1.41
H2-M-BR1	29.91	66.58	0.13	29.91	31.37	31.84	30.88	0.5	1	1.99	76	Warm	1.40	1.39
H2-M-BR2	30.43	69.14	0.17	30.43	31.91	32.52	31.48	0.5	1	2.07	80	Warm	1.44	1.45
H3-S-MR	30.75	71.43	0.09	30.75	32.23	32.55	31.65	0.5	1	2.38	90	Warm	1.59	1.67
H3-S-SR	30.93	70.52	0.17	30.93	32.42	33.05	31.99	0.5	1	2.18	84	Warm	1.50	1.53

Equations and sources
TBS = Dry bulb temperature (°C) = Ta = Indoor air temperature (°C)
TGN = Black globe temperature (°C), V = Indoor air velocity (m/s)
TGN = 0.456 + 1.033TBS (de Abreu et al. 2011)
Mean radiant (°C) = TGN + 2.42 V (TGN − Ta)
PMV/PPD: (Tyler et al. 2017)
aPMV = PMV/(1 + 0.21 * PMV): (Yao et al. 2009; De Dear and Brager 1998; Chongqing University, Chinese Academy of Architectural Sciences 2012)
PMVe = PMV * e: e = 0.7: (Zhu et al. 2008)

Table 5.7 Average values of the thermal comfort indices (PMV and PPD) and the averages values of PMV modifications for the three selected historic houses (H1, H2, H3)

House	PMV average	PPD average (%)	aPMV average	PMVe average
House 1	1.99	76	1.40	1.39
House 2	2.03	78	1.42	1.42
House 3	2.28	87	1.54	1.60

healthy environment going in line with energy efficiency for the historic buildings and encourage the continuity of their use adaptively in line with their structures, two modifications of the Predicted Mean Vote (PMV) have been calculated. First one is PMVe of (Fanger and Toftum 2001) with middle expectancy factor of rooms without air conditioning in the warm climate of 0.7 value. Second modification based on the Chinese Adaptive Predicted Mean Vote (aPMV) model after simulating Tartous climate with the Chinese mild climate zone, then using its adaptive coefficient value ($\lambda = 0.21$) in the calculating (Yao et al. 2009; De Dear and Brager 1998; Chongqing University, Chinese Academy of Architectural Sciences 2012). Table 5.7 shows the average values of the thermal comfort indices (PMV and PPD) and the averages values of PMV modifications for the three historic houses. The result obtained after calculating PMV index varies between 1.99 and 2.28, and PPD ranged from 76 to 87% of dissatisfied people. All of these houses classified into the warm category and have the value of PMV falls outside the range of the neutral (−1, 0, +1) with less than 34% of satisfied people that did not comply with related PMV international standards. The aPMV average varies from 1.4 to 1.54 similar to the PMVe average which varies from 1.39 to 1.6. The PMV modification values are still in the unacceptable category of the indoor thermal environment.

④ Comparative Analysis of PMV and AMV

By using regression analysis, the correlation between the Actual Mean Vote (AMV/TSV) and indoor operative temperature (OP) in summer has been obtained in addition to the correlation between the Predicted Mean Vote (PMV) and the indoor operating temperature (OP) in summer, as shown in Fig. 5.27 and Eqs. (5.2 and 5.3):

$$\text{AMV} = 0.4223\ \text{OP} - 11.155 \quad R^2 = 0.782 \tag{5.2}$$

When AMV = 0:

The actual indoor neutral operative temperature = OP (actual) = 26.42 °C in summer

$$\text{PMV} = 0.2123\ \text{OP} - 4.5675 \quad R^2 = 0.8703 \tag{5.3}$$

When PMV = 0:

The predicted indoor neutral operative temperature = OP (predicted) = 21.5134 °C in summer. The residents start actually to feel hot at 28.78 °C, and the actual thermal neutral temperature in the old city of Tartous is higher than that

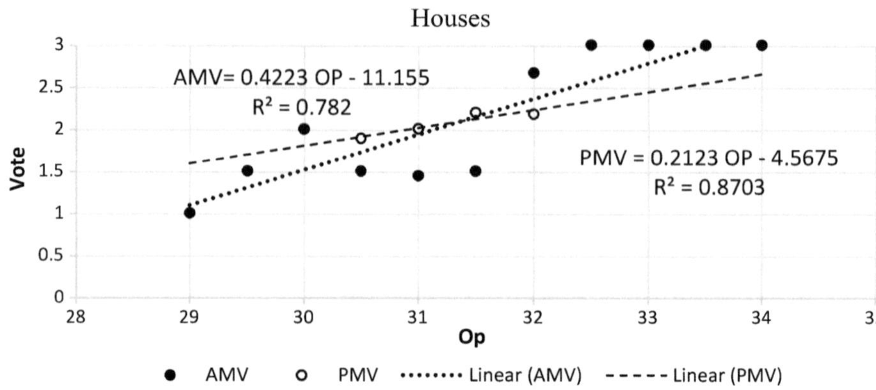

Fig. 5.27 Correlation of PMV and AMV with the Operative Temperature (OP) in the surveyed houses of the old city of Tartous in summer

derived from the PMV model. This indicates that in summer the residents in the old city of Tartous are more tolerant of the "hot" conditions. This may be explained by the local residents' action and adaptive measure that they use to improve their thermal comfort as the indoor air temperature increases. This requires finding the local standards and the local adjustment formula in further research.

⑤ Results' Comparison with the Previous Studies and International Standards

(1) The comparison of the obtained indoor parameters in both subjective and objective surveys with ASHRAE Standard 55-1992 shows that none of the studied samples falls within the required limits or the comfort band of the hot period, as shown in Table 5.8. Most of the surveyed dwellings had low ventilation rates and need an increase of ventilation ranging between 0.06 and 0.1 m/s. Despite the results showed the acceptance majority of the air movement in the surveyed dwellings of the old city of Tartous, a total 84.5% of the residents preferred to increase the air velocity because of the high relative humidity.

(2) A neutral temperature of 26.42 °C and a comfort band of 24.05 and 28.78 °C coinciding with −1 and +1 sensation vote have been obtained. As the indoor air temperature varied, the respondents' thermal sensation varied consequently and even their clothing varied. The thermal comfort zone found to be away from the top value of the range of temperatures (23–26 °C) specified by ASHRAE Standard −55. The residence is tolerant for a wide range of temperature in summer. Also, the study reveals that a majority of the occupants were willing to use passive cooling techniques. Hence, the proper strategies' implementation can result in significant reductions in energy consumption in our today's modern life.

(3) Based on ASHRAE standards, it can be drawn that the majority of the respondents (86.17%) in the subjective survey voted between "warm and hot sensation." The calculation of PMV–PPD model of the field measurement was in

Table 5.8 Subjective and objective results of the indoor parameters and thermal indices for the residential buildings

Indoor parameters/buildings	Air temperature (°C)	Relative humidity (%)	Airspeed average (m/s)	Operative temperature (°C)	Survey results		
					Comfort band	Neutral temperature (°C)	Category
Old city of Tartous	28.7–33.6	54.4–75	0.04–0.61	32.16	24.05−28.78	26.42	Warm and hot sensation
					PMV band	**PPD (%)**	**Category**
H1	28.2–31.3	61.6–80.7	0.07–0.34	30.87	−1.99 < PMV < + 1.99	76	Warm
H2	27.7–31.4	57.6–74.8	0.04–0.66	31.14	−2.03 < PMV < + 2.03	78	Warm
H3	30.1–31.5	66.2–75.2	0.05–0.32	31.82	−2.28 < PMV < + 2.28	87	Warm

the range of warm category, and more than two-thirds of the respondents were predicted to be not thermally comfortable with the indoor environment within their buildings. As well as the PMV modifications were in the unacceptable indoor environment category. Therefore, the environment in the dwellings in the old city of Tartous in summer is not within the recommended international acceptable thermal environment standards. In the meanwhile, the actual thermal tolerance of local people emphasized the need for finding local predicted standard based on the local climate.

(4) A clothing adaptation with the temperature was found in the dwellings of the old city of Tartous. The clothing insulation value was in broadly of about 0.50 Clo for male and 0.48 Clo for female. The values are near 0.5 Clo suggested for the hot period in the ASHRAE Standard 55-2004. The clothing adaptation was related to the occupants comfort more than sociocultural and economic aspects.

(5) Talking about the rate of the overall acceptability, 50% were voted to accept their houses' condition and 50% found their thermal environment unacceptable. However, as thermal acceptability is related to psychological factors and more easily biased than the thermal sensation votes, the method using thermal sensation votes is widely accepted more than the thermal acceptability one (McIntyre 1980).

(6) Considering the thermal preference scale, it is clearly shown that most respondents (86.2%) preferred to be a little bit cooler in their buildings. A 5.2% of the subjects indicated that changes in the thermal environment are not required, while 8.6% of the occupants they prefer to be too cool in order to overcome the indoor environmental condition. These votes related well to the overall acceptability vote, whereas no one stated that they prefer to be warmer as Tartous has a humid summer climate, so definitely, nobody would want to be warmer on summer period. In general, the results indicate that neutral thermal sensations are not always the preferred thermal state.

5.3 Microclimate Field Survey of the Selected Public Historic Buildings

5.3.1 Questionnaire Survey Results

① Basic Information

Most of the employees live in the city of Tartous, and only 10% of them travel to work every day from the nearby villages. All the survey participants in the (T) building worked on the first floor, and 90% of them performed professional activities related to cultural heritage. In the (D) building, 72.4% of the survey participants worked on the first floor, and 27.6% of them worked on the second floor; additionally, 80% of the total participants perform professional activities related to cultural heritage. The

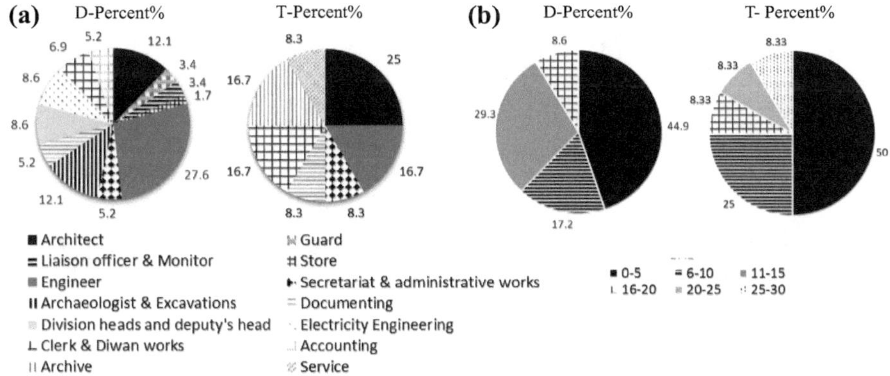

Fig. 5.28 **a** Respondents' occupation, **b** work duration of the respondents in the office buildings (D) and (T)

gender distribution of the participants was 30 males (51.7%) and 28 (48.3%) females in building (D) and nine females (75%) and three males (25%) in building (T). Most respondents had a high level of education. More than 50% of the respondents had worked in the same place for more than five years, as shown in Fig. 5.28.

② Indoor Environment Parameters Assessment

(1) Indoor air temperature and relative humidity

The measured values of the indoor parameters during the questionnaire survey for the main monitored structures are shown in Tables 5.9 and 5.10. The operative temperature (OP) was calculated according to ASHRAE Standard-55 (ASHRAE, ANSI 2004). The occupants were asked about their feelings in terms of humidity with a seven-point sensation scale, namely "Humidity Predicted Vote". Only 16.7% out of 12 respondents in building (T) and 50% out of 58 respondents in building (D) voted within the central three categories, showing that the occupants were not in acceptable thermal conditions within their residences according to the ASHRAE standard (ASHRAE, ANSI 2004), Fig. 5.29. A total of 83.3% of the occupants in building (T) overall voted that they felt uncomfortable and found the workplace to

Table 5.9 Measured values of the indoor parameters for the main measured structures

Occupant location in area	The average of the inside air temperature (°C)	The average of the relative humidity (%)	The average of the inside airspeed (m/s)	The average of the radiant temperature (°C)	The average of the calculated operative temperature (°C)
D	30.92	0.26	69.07	33.35	32.14
T	30.42	0.05	72.02	32.06	31.24

Table 5.10 Measured values of the indoor air temperature, relative humidity, wind speed, and the calculated operative temperature of the individual surveyed rooms

Building	Room	Total Number of responds	Males	Females	The average of the Inside Air Temperature (°C)	The average of the Inside Airspeed (m/s)	The average of the Humidity (%)	The Mean Radiant Temperature (°C)	The calculated Operative Temperature (OP)
Department of antiquities	ER	14	7	7	30.14	0.62	72.24	33.80	31.97
	TR	10	7	3	30.43	0.15	69.58	32.44	31.44
	Prospecting	18	3	4	31.7	0.08	72.3	33.51	32.61
	DIR	5	2	3	30.8	0.45	60	33.91	32.35
	Info Room	2	1	0	31.7	0.06	72.5	33.44	32.57
	Accounting	3	0	3	31	0.38	71	33.87	32.43
	DR	1	1	0	30.77	0.58	69.85	34.34	32.56
	Control room	4	4	0	30.8	0.25	71.85	33.19	31.99
	BR	12	3	6	30.5	0.3	71.73	33.05	31.78
	HBR	2	1	0	30.4	0.1	72.3	32.23	31.32
	Store	3	1	1	32.7	0.04	63.2	34.40	33.55
	Archive	2	0	1	30.1	0.14	62.3	32.06	31.08
Technical building	T-F1-DR	3	0	1	31.12	0.04	72.6	32.76	31.94
	T- F1-ODR	4	1	4	30.44	0.02	73.77	31.99	31.21
	T-F1-DocR	4	2	1	30.14	0.06	70.55	31.82	30.98
	T-F1-AR	2	0	4	30.5	0.02	74	32.05	31.27
	T-F2-ADocR	6	0	0	29.88	0.1	69.17	31.69	30.78

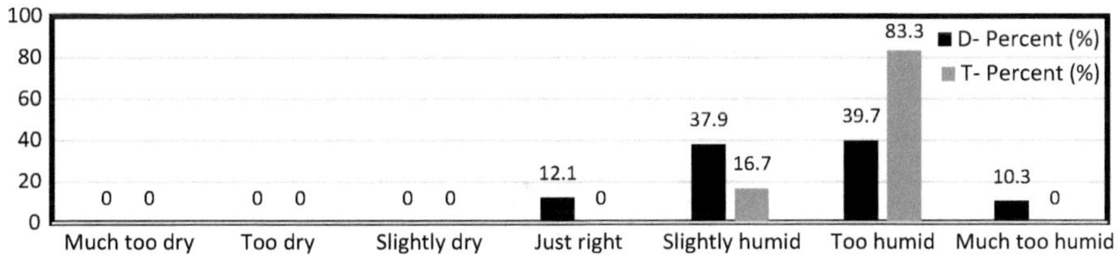

Fig. 5.29 Indoor Humidity Perception Vote (HPV) result

be too humid, and half of the occupants in building (D) overall voted that they were satisfied/comfortable with the indoor relative humidity. A total of 39.7% felt too humid, and the remaining 10.3% felt much too humid. This result indicates that the occupants were very sensitive to humidity variations.

(2) **Air velocity**

Figure 5.30 shows the Draft Perception Vote (DPV) results. Only 16.7% of the occupants in (T) felt the velocity was much too still. A total of 1.7% of the occupants in (D) felt the velocity was too still, and 6.9% of them felt that the velocity was too high. A total of 83.3% of the occupants in (T) and 91.4% of the occupants in (D) voted within the central three categories, which shows the acceptance of the air movement in their buildings. All the occupants of (T) preferred the airflow to be higher than present, 50% of the occupants in (D) preferred the airflow to remain at the present rate, 46.6% of them preferred it to be higher than present, and 3.4% of them preferred it to be lower than present.

(3) **Clothing insulation**

Male and female occupants in both office buildings wore a relatively light formal type of clothing. Their outfits provided them more comfort and made them feel cooler. The male clothing consisted of light trousers with a short- or long-sleeve shirt and differed in small ways such as underwear and footwear. The males' outfits, as calculated according to ASHRAE55-2004, ranged between 0.42 and 0.5 Clo units in both office buildings. The female clothing consisted of light trousers with a short- or long-sleeve shirt/blouse and differed in small ways such as a head cover, underwear, and footwear. The values associated with this clothing ranged between 0.4 and 0.54 Clo units in (D) and 0.4 and 0.57 Clo units in (T), as calculated according to ASHRAE55-2004.

(4) **Personal adjustments**

A summary of the answers to the multiple response questions about the personal response to hot conditions is shown in Fig. 5.31. The reactions of the employees to a sense of being hot did not require the use of a fan or cooler, and this reaction was not generally the sole response. Adjusting the indoor ventilation by opening the doors

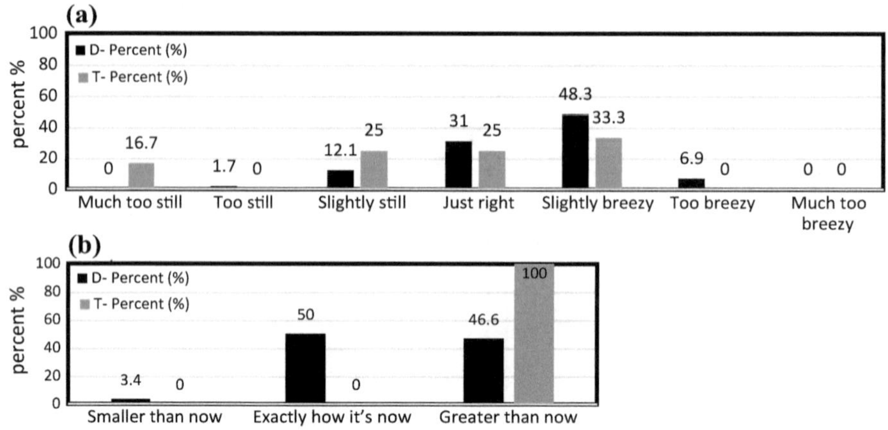

Fig. 5.30 Air velocity vote results in the office buildings: **a** indoor Draft Perception Vote (DPV) results, **b** the airflow preference

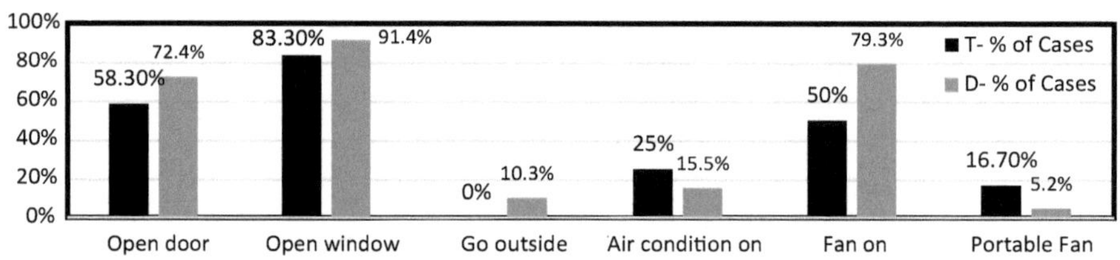

Fig. 5.31 Occupants adjustment/control choice

and windows and altering activity levels were the common responses. The variety of passive reactions formed a positive factor in terms of energy efficiency.

③ Thermal Response Vote Evaluation

(1) Thermal preferences

The results of the subjective thermal preference among the respondents are presented in Fig. 5.32. A high percentage of the respondents preferred slightly cooler conditions than the current indoor temperature in their buildings by a percentage of 83.3% in building (T) and 65.52% in building (D). The rest of the respondents, 16.7% of building (T) and 29.32% of the respondents in building (D), indicated that changes in the thermal environment were not required. A negligible percentage of the respondents (1.72%) wanted a slightly warmer indoor temperature, and approximately 3.44% wanted a too cool and much too cool indoor temperature in building (D).

(2) Thermal Sensation Votes (TSV)

The results of the subjective responses to temperature (thermal sensation) are presented in Fig. 5.33. The results show that the majority of the respondents of building

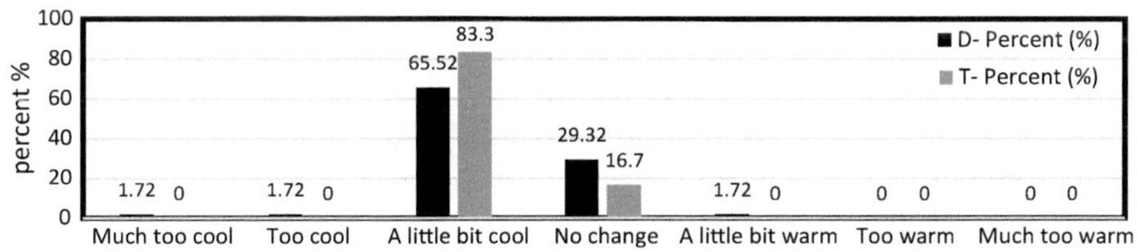

Fig. 5.32 Thermal preference vote results

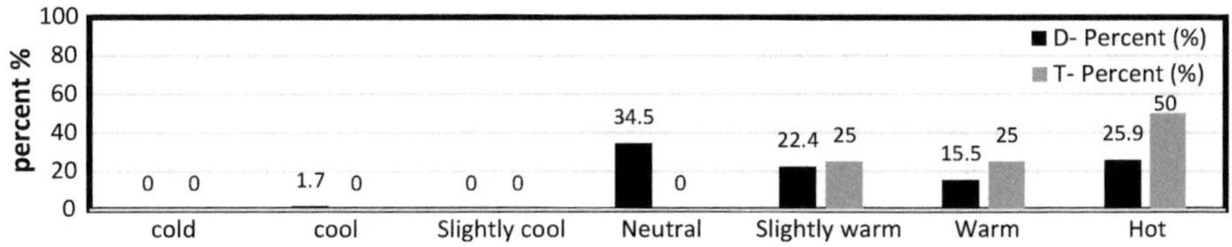

Fig. 5.33 Thermal Sensation Vote (TSV) results

(T) (75%) voted between "**warm and hot sensation**" and preferred cooler conditions, and 25% found their environment slightly warm. More than half of the respondents of the building (D) (56.9%) voted between "**neutral and slightly warm sensation**", and 41.4% voted between "warm and hot sensation". The occupants of the building (D) showed more tolerance for hot conditions. The comfort band of both office buildings was not calculated due to the group size after using BIN methods for the thermal sensation for an interval of 0.5 °C of the operative temperature (three variables for building T, and four variables for building D) which cause a narrow range of temperatures for the comfort band.

(3) Thermal acceptability

The results of the subjective thermal acceptability in the selected office buildings are presented in Fig. 5.34. All the participants of building (T) reported their thermal sensation as not acceptable; half of the respondents reported that the reason for their dissatisfaction was due to the temperature being always too hot, and the other half found it often too hot. On the other hand, 72.4% of the participants in building (D) accepted the overall temperature of their building, and 27.6% reported their thermal sensation as not acceptable because they found it occasionally too hot. According to the survey, the discomfort time occurred during the mid-day and afternoon (between 11 am and 5 pm) in building (D) and at all times in building (T).

④ Staff Comments and Views of the Thermal Comfort term and their Workplace Environment

The employees were asked about comfortable thermal conditions from their point of view, and their answers are shown below. The respondents in building (T) answered "feeling comfortable in the office during work time." The respondents in building (D) thought that the term "comfortable thermal conditions" means the following: "A

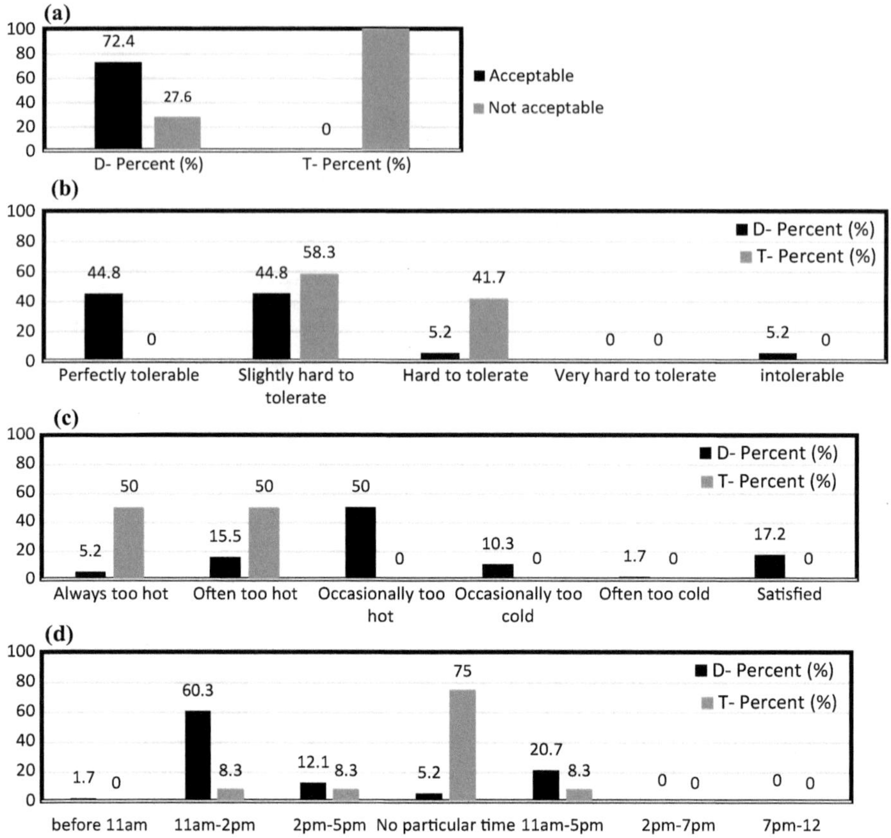

Fig. 5.34 Thermal acceptability and the discomfort time of the selected office buildings: **a** the acceptability vote results, **b** acceptability level, **c** the reason for thermal un/acceptability results, **d** the discomfort time in the selected office buildings

sense of comfort and satisfaction with the thermal status of the building;" "The indoor temperature is not very cold and not intensely hot, just a moderate temperature to suit the body;" "The satisfaction of indoor thermal conditions," "Optimal temperature of the body;" "Moderate microclimate without humidity," "No high moisture and no high temperatures;" "Moderate atmosphere and a suitable temperature for work;" "Adjust the temperature to an acceptable level, which lets us adapt to the surrounding environment;" "The sense of the surrounding environment temperature, both in terms of comfort or discomfort;" "Work in a place with a moderate temperature and low humidity;" "Necessary state is always required to get the best results in daily life and office work;" "Securing the heat that does not bother employees to do their work;" "Feeling comfortable as a result of suitable thermal conditions;" and "the status of thermal stability." On the other hand, the occupants in the office building commented in the open comments about their offices' environment and stated that the high temperatures in the building without an air conditioner are uncontrollable in the summer in building (T). Whereas most (D) building occupants accept the environment in summer due to the western direction of the rooms, which provides a nice atmosphere until 12:00 pm, later on, there is more glare and radiation. Approximately 70% of them mentioned the high air humidity in summer, especially in the absence of airflow that affects the electrical devices. The atmosphere of the rooms in building (D) in winter was cold and, in their opinions, needed thermal energy sources such as an oil fireplace.

5.3.2 Field Measurement Results

The measurement results of the indoor air temperature, indoor relative humidity, daily average of the indoor and outdoor wind speed, and indoor and outdoor air temperature are shown in Figs. 5.35, 5.36, 5.37 and 5.38, respectively. The maximum and minimum indoor air temperature and humidity in each measured room in the selected buildings are shown in Table 5.11. The indoor air temperature in the occupied zone for the Technical Office of the old city of Tartous (T) was varied between 27.6 °C and 31.6 °C and the indoor relative humidity was between 52.7 and 84.1%. The indoor air temperature as measured during the field survey in the Department of Antiquities in Tartous (D) was varied between 28.4 and 32.9 °C and the range of indoor relative humidity in the house two was within 49.7 and 72.4%. The indoor air temperature as measured in the public multifunctional gallery (G) was varied between 26.4 and 30.5 °C and the range of indoor relative humidity in it was within 74.6 and 96.2%. As well the indoor air temperature as measured in the print shop (P) was varied between 29.2 and 31 °C, and the range of indoor relative humidity was within 68.1 and 78.4%.

Four kinds of thermal behavior for the rooms could be distinguished in proportion to the construction periods. High relative humidity and a slight difference in the indoor air temperature and relative humidity between the nighttime and daytime can be seen at the measured points of the Crusader public multifunctional gallery, and the difference increases at the points near the openings of the waterfront facade, as shown in HP1. The indoor relative humidity values decreased by more than 20%, and the indoor air temperature values also increased more than 2 °C in the rooms built during other periods because of the different thermal masses. Relatively stable indoor air temperatures and relative humidity values and a slight difference between the nighttime and daytime can be seen in the cross-vaulted rooms, which belong to the Ottoman period in DIR and DR of the Department of Antiquities, in the DR, ODR, and DocR of the Technical Office, and in the print shop. A difference in the indoor air temperatures of approximately 3 °C between night and daytime can be seen in the rooms of the late Ottoman period ER and TR. In addition, there is an obvious difference in the indoor air temperature between nighttime and daytime in the rooms that were built in the French period. The temperature difference reaches

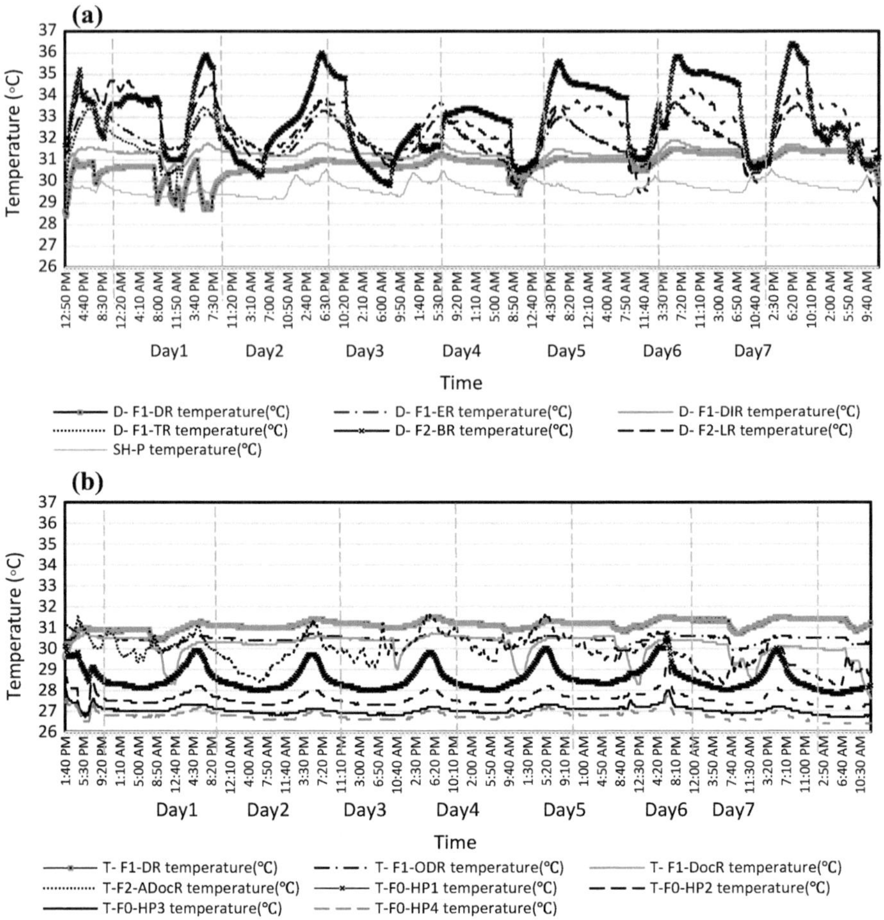

Fig. 5.35 Indoor air temperatures: **a** the rooms of the building (D) and the print shop, **b** the rooms of the building (T) and the gallery (G)

6 °C in BR on the second floor of the Department of Antiquities and approximately 3.5 °C in the ADocR on the second floor of the Technical Office.

The values of the indoor wind flow were generally slight values and ranged between 0.04 and 0.08 m/s in the first floor of the Technical Office (T) and the public multifunctional gallery (G) of the old city of Tartous. On the other hand, the values of the indoor wind flow in the rooms of the Department of Antiquities (D) in Tartous had high values and ranged between 0.16 and 0.67 m/s, and the average in the print shop (P) was 0.15 m/s. Furthermore, the second-floor room of the Technical Office (ADocR) had an average value of approximately 0.55 m/s, as shown in

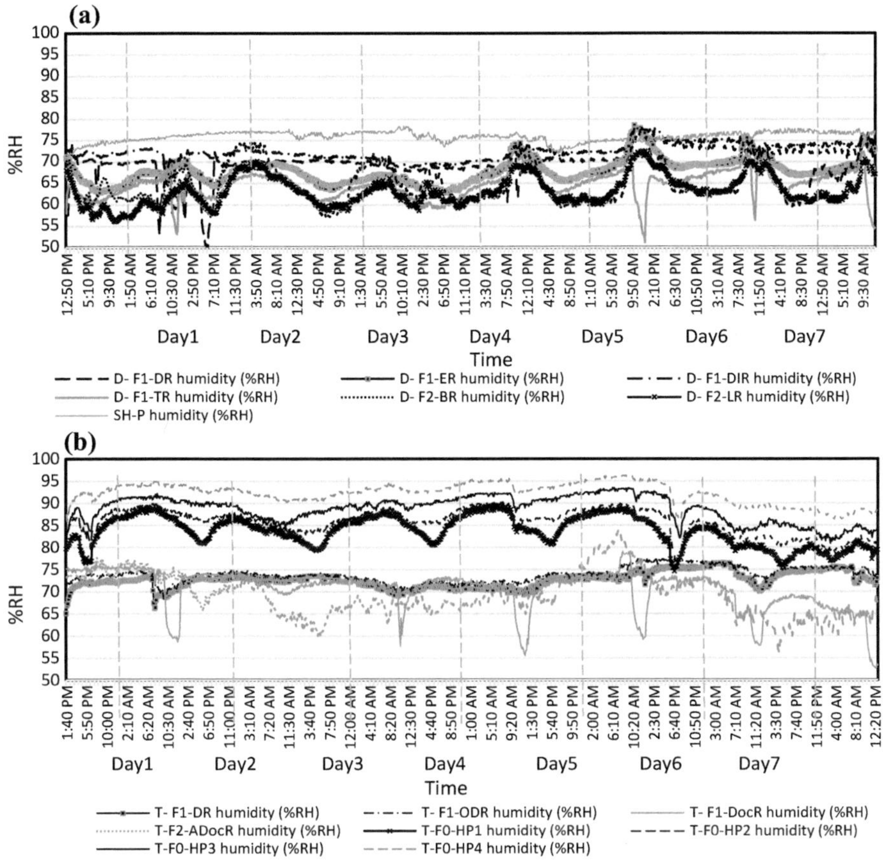

Fig. 5.36 Relative humidity: **a** the rooms of building (D) and the print shop, **b** the rooms of the building (T) and the gallery (G)

Fig. 5.37. The indoor air temperature values in the measured rooms on the first floor of the Department of Antiquities (D) and the print shop (P) were lower than the outdoor air temperature by approximately 0.5–1 °C. However, the indoor temperature values were more than the outdoor air temperature by approximately 0.5–1 °C in the measured rooms on the second floor of the Department of Antiquities (D). The natural ventilation played a positive role in decreasing the indoor air temperature in the second-floor room (ADocR) in the Technical Office. The lack of ventilation in its first-floor rooms (DR, ODR, DocR) had a negative impact by increasing the indoor air temperature to be more or approximately equal to the outdoor air temperature, as shown in Fig. 5.38. The thermal mass of the public gallery (G) played a significant role in decreasing the indoor air temperature values by approximately 2 °C less than the outdoor temperatures.

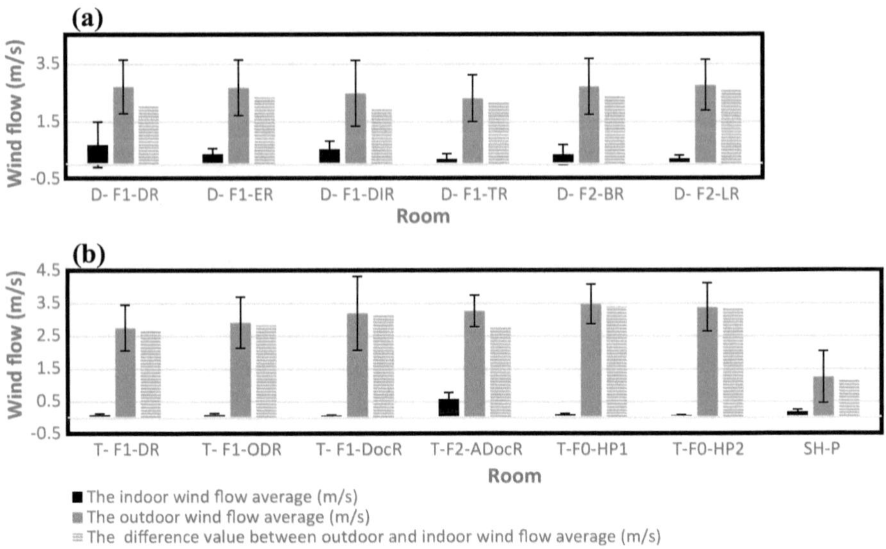

Fig. 5.37 August daily average of indoor, outdoor and the difference of the wind flow for the rooms of **a** the building (D), **b** the building (T), the gallery (G) and the print shop

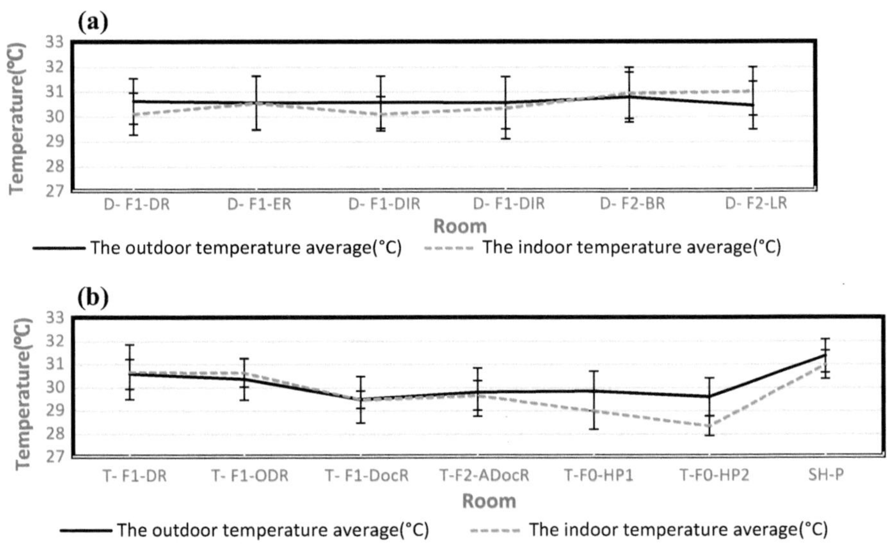

Note: vertical error bar refers to the standard deviation

Fig. 5.38 August daily average of indoor, outdoor and the difference of the temperature for the rooms of **a** the building (D), **b** the building (T), the gallery (G), and the print shop

① Variation in the Air Temperature and Humidity over the Measurement Intervals

The averages of the indoor air temperature and humidity in each day periods (intervals of times) of the all measured rooms and points for the Department of Antiquities in Tartous (D) and the Technical Office of the old city of Tartous (T) and the public multifunctional gallery (G) were configured in Fig. 5.39. The measured cross-vaulted rooms of the Department of Antiquities (D) showed stable air temperature and relative humidity values during the daytime. The other measured rooms of building (D) showed an increase in the air temperature (approximately 3 °C) and a decrease in the relative humidity (approximately 6%) after working hours began from 2 pm to 11 pm, especially in the second-floor rooms. Therefore, their indoor space at morning and noon was better than during the daytime. The second-floor rooms had the highest temperature, especially in the afternoon and evening hours, because of the outdoor solar radiation on both the ceiling and western facades.

The measured rooms on the first floor of the Technical Office of the old city of Tartous (T) showed relatively similar averages of the air temperature and relative humidity during the daytime intervals. Although there was solar radiation on both the ceiling and the western facades of the second-floor room during the daytime, it had a relatively low temperature compared to the first-floor rooms. This indicates that the ventilation there played a role in reducing the indoor air temperature. The gallery (G) had a relatively steady indoor air temperature and relative humidity during the daytime intervals compared with the other measured rooms in the public buildings that were built in different periods and with different building envelopes. It had the lowest air temperature, with an average of 27.5 °C, and the highest relative humidity, with an average 85.4%, during the measurement period. The print shop (P) had a relatively steady average indoor air temperature (between 29 and 30 °C) and relative humidity (between 75 and 77%) during the daytime intervals of the measurement period.

② Variation in the Air Temperature and Relative Humidity of the Selected Public Buildings for Each Measured Day

All the selected public buildings have western orientation and are built of sandstone, which may explain the reason for the parallel lines of the indoor parameter averages for each measured day, as shown in Fig. 5.40. The relative humidity was inversely proportional to the indoor air temperature in all the selected buildings. The public gallery (G) showed the highest averages of relative humidity and the lowest averages of indoor air temperature during the measurement period compared with the other measured spaces. The building (T) and the print shop (P) had high humidity levels in sync with the high indoor temperatures, which lead to an inefficient evaporative cooling of the skin that caused discomfort. Despite having the highest average temperature values, the indoor relative humidity average values in building (D) were better than those of the other buildings, which led to more human comfort inside the rooms.

Table 5.11 Maximum and minimum indoor air temperature and humidity in each measured room in the selected public buildings

Building	Technical Office of the old city of Tartous (T)				Department of antiquities in Tartous (D)						Public multifunctional gallery (G)				Print shop
Floor	T- F1			T-F2	D-F2				D-F1		H-F0				P-F0
Room name	DR	SR	DocR	ADocR	DR	ER	DIR	TR	BR	LR	HP1	HP2	HP3	HP4	P
Temp Av (°C)	31.12	30.44	30.14	29.88	31.5	33.7	31.9	33.8	36.4	34.7	28.58	27.59	27.04	26.77	29.70
Temp max (°C)	31.5	30.8	30.7	31.6	30.8	32	31.3	32	32.9	32.7	30.5	29.4	28.4	27.7	31
Temp min (°C)	29.6	29.8	27.6	28.2	28.4	29.8	30.2	28.6	29.8	30.6	27.8	26.7	26.7	26.4	29.2
Rh Av (%)	72.60	73.77	70.55	69.17	77.6	78.6	78.3	74.1	77.2	72.4	83.75	85.77	88.67	92.24	75.9
Rh max (%)	76.5	77.3	79.1	84.1	69.9	67.9	72.4	64.6	65.3	63.3	89.5	90.3	93.4	96.2	78.4
Rh min (%)	65.1	68.3	52.7	55.8	49.7	62.8	68.7	51.2	57.2	56	74.6	75.7	80.9	85.9	68.1

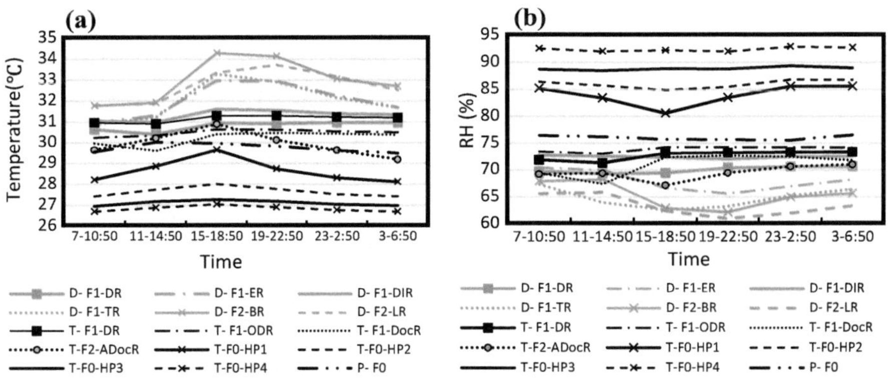

Fig. 5.39 Input results during the daytime intervals of all the measured rooms and points of the selected public buildings: **a** air temperature, **b** relative humidity

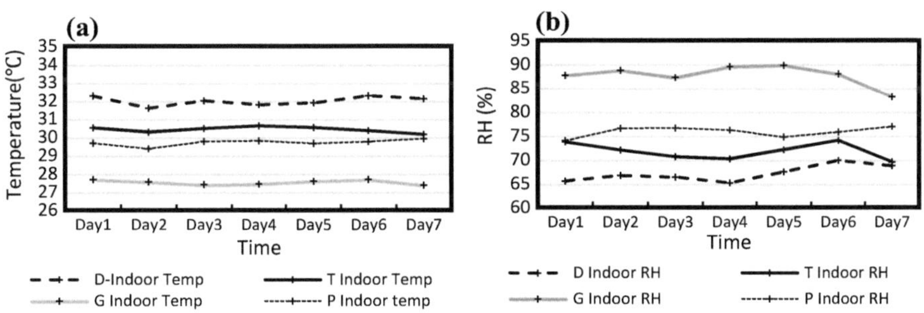

Fig. 5.40 Variation of indoor average air temperature and relative humidity of the selected public buildings: **a** indoor air temperature, **b** indoor relative humidity

③ Thermal Comfort Indices

The calculations of the thermal comfort indices, the Predicted Mean Vote (PMV) and the Predicted Percentage of Dissatisfied (PPD), and the modifications (aPMV, and PMVe) of the Predicted Mean Vote (PMV) for the selected public structures (D, T, G, and P) were performed in the same equations and tool that were used in the study of residential buildings, as they are shown in details in Table 5.12. Table 5.13 shows the average values of the thermal comfort indices (PMV and PPD), and the averages values of PMV modifications for all the case studies. The PMV index for building (D) varied from 2.71 for the first floor to 4.21 for the second floor, with a mean value of 3.21 for all the building rooms. For building (T), the range was from 1.15 for the second floor to 2.35 for the first floor, with a mean value of 2.05. The average of PMV of the gallery (G) was 1.23, and for the print shop (P) was 1.95. By equating to the ASHRAE thermal sensation scale (−3 cold, −2 cool, −1 slightly cool, 0 neutral, +1 slightly warm, +2 warm, +3 hot), the results show that building (D) was observed to be in the range of the "hot" category with average PPD 93.33%. Building (T) was in

Table 5.12 Results of calculating the thermal comfort indices (PMV and PPD), PMV modifications for the selected public historic structures (D, T, G, P)

Room name	Indoor air temperature (°C)	Indoor relative humidity (%)	Indoor air velocity (m/s)	TBS = Dry bulb temperature (°C)	TGN = Black globe temperature (°C)	Mean radiant (°C)	Operative temperature (°C)	Clothing (Clo)	Metabolic Rate (met)	PMV	PPD (%)	Sensation	aPMV	PMVe
D-F1-DR	31.50	77.60	0.67	31.50	33.01	35.46	33.48	0.5	1	1.94	74	warm	1.38	1.36
D-F1-ER	33.70	78.60	0.33	33.70	35.28	36.55	35.13	0.5	1	3.08	99	hot	1.87	2.16
D-F1-DIR	31.90	78.30	0.49	31.90	33.42	35.23	33.57	0.5	1	2.25	87	warm	1.53	1.58
D-F1-TR	33.80	74.10	0.16	33.80	35.39	36	34.9	0.5	1	3.58	100	hot	2.04	2.51
D-F2-BR	36.40	77.20	0.31	36.40	38.08	39.33	37.87	0.5	1	4.50	100	hot	2.31	3.15
D-F2-LR	34.70	72.40	0.17	34.70	36.32	36.98	35.84	0.5	1	3.91	100	hot	2.15	2.74
T-F1-DR	31.12	72.60	0.07	31.12	32.62	32.87	32	0.5	1	2.57	95	hot	1.67	1.80
T-F1-ODR	30.44	73.77	0.07	30.44	31.91	32.16	31.3	0.5	1	2.31	89	warm	1.56	1.62
T- F1-DocR	30.14	70.55	0.06	30.14	31.60	31.82	30.98	0.5	1	2.17	84	warm	1.49	1.52
T-F2-ADocR	29.88	69.17	0.55	29.88	31.33	33.27	31.58	0.5	1	1.15	33	Slightly Warm	0.93	0.81

(continued)

Table 5.12 (continued)

Room name	Indoor air temperature (°C)	Indoor relative humidity (%)	Indoor air velocity (m/s)	TBS = Dry bulb temperature (°C)	TGN = Black globe temperature (°C)	Mean radiant (°C)	Operative temperature (°C)	Clothing (Clo)	Metabolic Rate (met)	PMV	PPD (%)	Sensation	aPMV	PMVe
T-F0-HP1	28.58	83.75	0.08	28.58	29.99	30.26	29.42	0.5	1	1.66	59	warm	1.23	1.16
T-F0-HP2	27.59	85.77	0.04	27.59	28.97	29.10	28.35	0.5	1	1.26	38	Slightly Warm	1.00	0.88
T-F0-HP3	27.04	88.67	0.03	27.04	28.40	28.40	27.72	0.5	1	1.04	28	Slightly Warm	0.85	0.73
T-F0-HP4	26.77	92.24	0.02	26.77	28.12	28.12	27.45	0.5	1	0.96	24	Slightly Warm	0.80	0.67
SH-P	29.70	75.90	0.15	29.70	31.16	31.69	30.7	0.5	1	1.95	74	warm	1.38	1.37

Table 5.13 Average values of the thermal comfort indices (PMV and PPD) and the averages values of PMV modifications for the selected public historic structures (D, T, G, P)

	PMV average	PPD average (%)	Sensation	aPMV average	PMVe average
Building (D)	3.21	93.33	Hot	1.88	2.25
Building (D)—First floor	2.71	90	Hot	1.71	1.9
Building (D)—Second floor	4.21	100	Hot	2.23	2.94
Building (T)	2.05	75.25	Warm	1.41	1.44
Building (T)—First floor	2.35	89.33	Warm	1.57	1.65
Building (T)—Second floor	1.15	33	Slightly Warm	0.93	0.81
Gallery (G)	1.23	37.25	Slightly Warm	0.97	0.86
SH-P	1.95	74	Warm	1.38	1.37

the range of the "warm" category with average PPD 75.25%. The public gallery (G) was "slightly warm", and average PPD 37.25% of dissatisfied users, and the print shop (P) average PPD 74% of dissatisfied users and classified in "warm" category. The case studies were expected to be dissatisfied (uncomfortable) with the thermal environment, where the required PPD value should be less than 27%, according to Chinese standards, and less than 10% according to ASHRAE standard. The PMV modifications values of the historic gallery were 0.97 for the aPMV, and 0.86 for the PMVe and that shows an acceptable expected naturally ventilated environment. The other case studies' PMV modification values showed unacceptable indoor thermal environment.

By comparing the objective and subjective results, the rooms in building (T) had a PPD of 75.25%, which was operating with unanimously unacceptable indoor environment complied with the results of the thermal preference of the questionnaire of the subjective survey, where (75%) of the respondents voted between "warm and hot sensation" due to the lack of ventilation. On the other hand, the predicted dissatisfaction of building (D), with PPD of 93.33%, did not comply with the actual occupant thermal acceptability and preference results of the questionnaire. However, approximately two-thirds of the participants in building (D) accepted the overall temperature of their building. More than half of the respondents in building (D) (56.9%) voted between "neutral and slightly warm sensation," and 41.4% voted between "warm and hot sensation." This requires finding the local standards and the local adjustment formula in further research.

① **Results' Comparison with the Previous Studies and International Standards**

No regulations specify the thermal standards for the workplace in Syria. The American Society of Heating, Refrigerating, and Air-Conditioning Engineers (ASHRAE) Standard 55-2010 recommends the temperature/humidity ranges for comfort for the summer period with light clothing. The ASHRAE's acceptable ranges of operative temperatures are 24.5–28 °C and 23–25.5 °C for the relative humidity levels of 30% and 60%, respectively (ASHRAE, ANSI 2010). A high relative humidity supports mold and bacterial growth and causes discomfort such as excessive perspiration and exacerbation of the effects of high temperature. A low relative humidity can cause respiratory problems. According to EN15251 (BSI 2008), increased airspeed using personal ventilation systems or fans may be applied to elevate the upper comfort temperature limit when indoor operative temperatures exceed 25 °C in summer. Many researchers have examined the comfort of ASHRAE-55 for their countries in recent decades. Although the results of thermal temperature in different studies are different, the common point is that either they are very close to the indoor air temperature or above the limit of the required indoor temperature. Additionally, the occupants of the historical and natural ventilation buildings were more tolerant of the thermal environment than the new building occupants. The subjective and objective results of our survey are shown in Table 5.14.

(1) The measured indoor air temperature and relative humidity compared with the ASHRAE 55-2004 range show that none of the studied samples were within the required limits. The occupants were sensitive to the humidity, particularly the occupants of the building (T).
(2) An increase in the indoor wind flow values between 0.07 and 0.11 m/s in the rooms of the first floor of the building (T) and the multifunctional public gallery (G) is needed. While the measured rooms of the building (D) and the second floor of the building (T) had proper ventilation.
(3) The Predicted Mean Vote (PMV) and the Predicted Percentage of Dissatisfied (PPD) in all four measured samples did not comply with the international and Chinese Standards. That emphasized the importance of more research for finding a local adapting formula and thermal standards for naturally ventilated and historic buildings.
(4) The clothing adaptation was a relatively light, formal type related to the clothing of common people near the value of 0.5 Clo suggested for the summer season in the ASHRAE Standard 55-2004.

5.4 Key Strategies and Recommendations

Preservation of the built heritage is an essential task for both societal and community well-being. The authenticity and special considerations need to be taken when suggesting the appropriate usage function of heritage buildings. Improvements considering the building envelope and value are helpful to adapt the indoor environment

Table 5.14 Subjective and objective results of the indoor parameters and thermal indices for the public buildings

Building	Subjective survey					Objective results				
	Air temperature (°C)	Relative humidity (%)	Airspeed average (m/s)	OP (°C)	Category	Air temperature (°C)	Relative humidity (%)	PMV band	PPD (%)	Category
D	30.1–32.7	60–72.5	0.26	32.14	Neutral and slightly warm	28.4–32.9	49.7–72.4	$-3.21 < PMV < +3.21$	93.33	Hot
T	29.88–31.12	69.17–74	0.05	31.24	Warm and Hot	27.6–31.6	52.7–84.1	$-2.05 < PMV < +2.05$	75.25	Warm
G	Not surveyed					26.4–30.5	74.6–96.2	$-1.23 < PMV < +1.23$	37.25	Slightly Warm
P	Not surveyed					29.2–31	68.1–78.4	$-1.95 < PMV < +1.95$	74	Warm

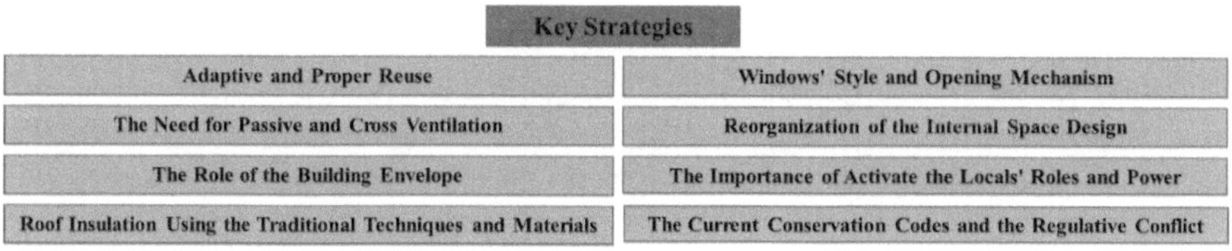

Fig. 5.41 Key strategies in the microlevel

for the occupants and to increase productivity during the operating hours. As this is the first research on thermal comfort in Tartous, these overall results indicate that it may prove beneficial to develop thermal comfort regulations in Tartous for historic residential buildings based on ASHRAE, ISO, and Chinese standards which would raise the occupant's satisfaction and their perceived performance as well as potential results in the decrease of energy consumption in the buildings and enhancing their sustainability. The key strategies could be useful for further studies and could be used as a dataset for establishing a local thermal comfort standard, Fig. 5.41.

① Adaptive and Proper Reuse

The transformation of the old city of Tartous from defensive usage to residential one formed a continuous historic development as a living heritage, where the traditional way of life became an integral part of the site's values and gave a different value. The purpose of using the historic buildings changed and the interior design of the buildings themselves also changed differently from the original design, such as dividing the same space into other small spaces and covering the courtyards and the outdoor yards of the houses. Even though the interior arrangement of the space changed, the shading devices and natural ventilation design remained as it was and perhaps worse. The proper and adaptive reuse should consider the structure envelope, occupants' needs, indoor conditions, and value. The adjustments and the added facilities must go in line with the modern needs based on the research indices with effective solution and without damage the authenticity or changing the values.

② The Need for Passive and Cross-Ventilation

The high registered humidity that was determined from the results means that the ability of the occupants for evaporative cooling has reduced significantly which leads to discomfort. Due to this high humidity, the current natural ventilation is ineffective and reduce the humidity would be in increasing the incoming air. The natural ventilation strategies using wind pressure, night and cross-ventilation are effective methods for providing the necessary air velocities needed to create comfortable indoor conditions at low costs. Application of openings between rooms such as windows and doors as well the casement windows are techniques to complete the airflow journey through a building, such applications in the building design if well integrated can provide low consumption of energy inside the buildings and improve thermal comfort condition.

③ The Role of the Building Envelope

The building envelope, the ventilation, and the construction period for each floor play a significant role in the thermal performance and occupant satisfaction. The construction period highlights the previous primary vision of the adaptive reuse and energy needs of the historic structures of similar periods. All the monitored public historic structures are built of sandstone and have the same building orientation, which enabled us to distinguish the difference between the thermal performance of each floor and construction period (Crusader, Ottoman, late Ottoman, and French period). Moreover, the thermal performance of the surveyed houses of the old city of Tartous which is mostly related to the late Ottoman and the French mandate periods and the crusader forts with the courtyard design in Arwad Island emphasized this result.

④ Roof Insulation Using the Traditional Techniques and Materials

Special attention should be given to the final floor rooms in all study cases. They have a higher temperature than the other floor rooms and furthermore higher than the outdoor temperatures in the noon and afternoon periods. Sun radiations are the primary source of heat gain, and the heat influences them from the roof and walls during the daytime and the relatively large amount of solar radiation from the openings. Enhancing the structure envelope's quality and installing the thermal insulation and shading would reduce heat from the external environment and help the residents to be thermally comfortable within their buildings. This study area is under conservation heritage site, the traditional material would be suitable to enhance the structure envelope's quality, such as using lime plaster to treat the building envelope and roof insulation.

⑤ Windows' Style and Opening Mechanism

The results of both field survey and investigation of the dwelling indicate that the current shading and natural ventilation strategies are not providing the desired comfort conditions that occupants seek, and two-thirds of the occupants are still dissatisfied with their thermal conditions. That caused an increase in the need and use of mechanical ventilation systems that would be resulted in increased cooling loads and an increase in the country energy intensity and consumption. Redesigning the fenestration or providing the sun shading tools taking into account the consideration of preservation to make a balance between the original envelope shape and the occupants' needs could improve the building envelope and thermal comfort.

⑥ Reorganization of the Internal Space Design

Though several buildings had a good design regarding other aspects and received better prevailing wind incident, they cannot thoroughly enjoy the benefits air circulation just because of poor internal space arrangement that restricts the cross-ventilation effect. Therefore, the arrangements of the interior space of room layout are essential to support the cross-ventilation effect in the house and increase the comfort inside the building. Reduce some of the unwanted or unnecessary furniture also will reduce the heat gain inside the space and improve the thermal comfort within the structure.

⑦ **The Importance of Activate the Locals' Roles and Power**

The passive reactions in the hot period of the locals were common responses in all the case study sites. Adjusting clothing, altering activity levels, opening the doors and windows form a positive factor regarding energy efficiency, and emphasize the possibility of increasing awareness about simple passive habits and practice to achieve passive strategies. More than half of the respondents of the old city of Tartous residents were of the young category, and this gives a further opportunity to work on raising awareness and changing their adjustment behavior. The civilians and professionals from government and other authorities can work together successfully as long they know and accept their role, on different scales and different time schedules.

⑧ **The Current Conservation Codes and the Regulative Conflict**

Thermal comfort in the indoor environment in the historic buildings is an important issue for the occupants' health and preserving the place identity and the continuity of the local architectural heritage and the cultural role of the old cities and maintains their human resources and historic value. Additionally, it has a direct influence on the energy usage. The municipality has a significant role in changing and implementing conservation strategies. The current conservation code and legislation of the old city of Tartous are strict. Therefore, the conflict between the related authority and the residents' needs should deeply discuss to modify the legislation and take the proper measures to accommodate the current usage, climatic variations, and thermal comfort.

The small sample size which is analyzed is still not enough to determine an absolute variable pattern of the desired thermal sensation in the Syrian coast. Further studies concerning thermal comfort should also be performed in suburban areas to obtain a wide conception of thermal comfort circumstances in the areas with different cultural and behavioral patterns. Future studies should entail a detailed examination of the effects of the parameters affecting thermal satisfaction through a computer simulation analysis and accompanied by a prototype construction for measurements. The recorded and documented traditional windows and shading devices in the old city of Tartous could also be examined to find their performance. That would provide more comprehensive study, especially in validating theoretical methods for thermal predictions of any restoration or future building design. The sustainability, natural ventilation strategies, and statistical thermal data collection need to be considered in framing the local standards. Such standards can have an effect on the locals' thermal comfort and the energy use of the structures, where acceptable low-energy solutions are available they should be preferred.

5.5 Summary

This chapter presented an overview of the old city of Tartous' urban fabric, public buildings, and dwellings. A thermal comfort survey of approximately 13.3% of the total population randomly covered the old city of Tartous, and field measurements for three historic houses have been conducted. Additionally, the results of a thermal comfort field measurement for four public historic structures in the old city of Tartous and a questionnaire for approximately 80% of the total number of staff in two office buildings during August 2016 were also presented. Demographic information and data about building envelope, architectural design, and materials data of the selected structures were collected. The main thermal parameters were measured and assessed. The thermal comfort band and PMV and PPD indices have been calculated with an analysis of the indoor climates. Finally, the results obtained in this study were compared with the previous studies and international standards to evaluate the thermal condition and the occupants' comfort and to find appropriate key strategies. Measurement results confirmed the importance of providing cross-ventilation and the need for reducing the relative humidity during the hot period. To improve the historic building envelope; using lime insulation for the roofs, changing the fenestration design, and providing the sun shutters in adapting to the context of the old city of Tartous were highly recommended. Further studies concerning the thermal comfort and the need for preservation legalization based on the comfort considerations and the needs of the local residents were highly recommended. The results of this survey could be used as a preliminary guide for the analysis of the retrofitting plans and materials for the historic buildings, and a dataset toward an approach to environmental design for the new buildings based on the local climate and thermal considerations.

References

Amer, R., & Abou Hamed, F. (2011). Optimal study design for Wind farm in Arwad Island. *Energy Procedia, 6,* 721–735.

Antiquities and Museums Department in Tartous. (2015). *Arwad Island's historic buildings*. Tartous, Syria: Antiquities and Museums Department, historic buildings section.

ASHRAE, ANSI. (1992). *Standard 55–1992, Thermal environmental conditions for human occupancy.* Atlanta, GA: American Society of Heating, Refrigerating and Air-Conditioning Engineers

ASHRAE, ANSI. (2004). *Standard 55–2004, Thermal environmental conditions for human occupancy.* Atlanta, GA: American Society of Heating, Refrigerating and Air-Conditioning Engineering

ASHRAE, ANSI. (2010). *ANSI/ASHRAE 55–2010 Thermal Environmental Conditions for Human Occupancy.* Atlanta, GA: American Society of Heating, Refrigerating and Air-Conditioning Engineering

Bashour, N. (2011). *Documentation of the old city of Tartous.* Damascus: Tartous Governorate in cooperation with the Municipal Management Modernization Program (MAM), funded by the European Union

Becker, R., & Paciuk, M. (2009). Thermal comfort in residential buildings—failure to predict by Standard model. *Building and Environment, 44,* 948–960.

Braune, M. (1985). *Die mittelalterlichen Befestigungen der Stadt Tortosa-Tartūs: Vorbericht der Untersuchungen 1981–1982. Damaszener Mitteilungen: Verlag Philipp von Zabern* (p. 2).

BSI. (2008). *BS EN 15251:2007, Indoor environmental input parameters for design and assessment of energy performance of buildings addressing indoor air quality, thermal environment, lighting and acoustics*. London: British Standards Institute.

Buratti, C., & Ricciardi, P. (2009). Adaptive analysis of thermal comfort in university classrooms: Correlation between experimental data and mathematical models. *Building and Environment, 44,* 674–687.

Chongqing University, Chinese Academy of Architectural Sciences. (2012). *Evaluation standard for indoor thermal environment in civil buildings GB/T 50785-2012 (In Chinese)* (pp. 11–12). Beijing: China Construction Industry Press.

de Abreu, P. G., et al. (2011). Estimativa da Temperatura de Globo Negro a Partir da Temperatura de Bulbo Seco. *ENGENHARIA NA AGRICULTURA/Engineering in Agriculture, 19*(6), 557–563.

De Dear, R. & Brager, G. S., (1998). Developing an adaptive model of thermal comfort and preference. *Ashrae Trans, 104*(1), 73–81(9)

Deschamps, P. (1973). *Les châteaux des croisés en Terre-Sainte III* (Vol. 19). Paris: P. Geuthner.

Fanger, P. O. & Toftum, J., (2001). Thermal comfort in the future-excellence and expectation. In *Conference on moving thermal comfort standards into the 21st century*. Oxford Brookes University, s.l., s.n

Hassan, M., & Rahmoun, T. (2016). *Protection strategy of the climate change for coastal heritage sites in Syria* (pp. 474–483). Nanjing: China, Digital Media press of Nanjing University.

Hassan, M., Xie, H., & Rahmoun, T. (2018). Field study in the residential buildings in the old city of Tartous in Syria during the summer period. In *Resilient & Responsible Architecture & Urbanism (RRAU) Conference.* Groningen, Netherlands: s.n

ISO7726. (1998). *International Standard ISO 7726. Ergonomics of the Thermal Environment, Instruments For Measuring Physical Quantities*. Geneva: International Standard Organization.

Karfoul, D., & Mouton, J.-M. (2008). *La conception architecturale de l'habitat urbain en orient meditarraneen du xiv au xix siecles-etude d architecture domestique-l exemple de la vieill ville Tartous, Syrie (Doctoral dissertation).* Lyon: Université Lumière Lyon 2

Kennedy, H. (1994). *Crusader castles*. Cambridge, New York: Cambridge University Press.

McIntyre, D. (1980). Design requirements for a comfortable environment. Bioengineering. In K. Cena & J. A. Clark, (Eds.), *Thermal Physiology and Comfort* (pp. 157–168). Amsterdam: Elsevier

Meteorological Center of Tartous. (2016). *Third and fourth periodic reports of August for the year 2015*. Tartous: Meteorological Center of Tartous, Syria.

Neglia, G. A. (2009). The Multicultural fabric of the crusader citadel of tartous in Syria. In A. Petruccioli & A. Sarro (Eds.), *Beyond the wall, notes an multicultural mediterranean landscape* (pp. 39–50). Bari, Italy: Uniongrafica Corcelli.

Nematchoua, M. K., Tchinda, R., & Orosa, J. A. (2014). Thermal comfort and energy consumption in modern versus traditional buildings in Cameroon: A questionnaire-based statistical study. *Appl Energy, 114,* 687–699.

Pillet, M. (1929). Notre-Dame de Tortose. *Syria. Archéologie, Art et histoire*, *10*(1), 40–51.

Rey, E. G. (1866). *Essai sur la domination française en Syrie durant le moyen âge*. Paris: Impr. E. Thunot.

Ricciardia, P., & Buratti, C. (2015). Thermal comfort in the Fraschini theatre (Pavia, Italy): Correlation between data from questionnaires, measurements, and mathematical model. *Energy and Buildings, 99,* 243–252.

Ricciardia, P., Zilettib, A., & Burattic, C. (2016). Evaluation of thermal comfort in an historical Italian opera theatre by the calculation of the neutral comfort temperature. *Building and Environment, 102,* 116–127.

Tyler, H. et al. (2017). *CBE Thermal Comfort Tool ASHRAE-55.* [Online] Available at: http://comfort.cbe.berkeley.edu/

Yao, R., Li, B., & Liu, J. (2009). A theoretical adaptive model of thermal comfort–Adaptive Predicted Mean Vote (aPMV). *Building and Environment, 44*(10), 2089–2096.

Zhu, Y., et al. (2008). Experimental Study on the Influence of Operating Temperature, Clothing Thermal Resistance and Season on Human Thermal Sensation in Partial Thermal Environment (In Chinese). *National Academic Collection of HVAC Refrigeration*

Chapter 6
Zoom Out–Zoom In Conservation Strategies and Intervention Levels

History is not just the significant past, it is evidence of a continuing commitment to tradition in the physical signifies the importance of heritage, and it is a symbol of cultural resilience. The historic structure is more than a physical thing; it differentiates the landscape. The place provides society with the opportunity to merge our memory of the past, challenges of the present, and hopes for the future. It allows us to contemplate essential things. It is necessary to begin developing the comprehensive plans for managing the threatened archaeological and natural sites that of the climate change according to the monitoring methodology, including the management stages. Each stage is evaluated using specific indicators whose values range from a starting scale to a critical situation to final decision making. Then examine the impact of management of these sites on the same property and the surrounding land and the community alike, to become mandatory management plans for protecting the built heritage in the era of climate change. This chapter presents Zoom Out and Zoom In approaches as mixed methods draw the frame of our research conservation principles and guidelines. Moreover, two conservation frameworks and processes in both macrolevel and microlevel for preserving the threatened coastal historic sites and buildings would be presented, considering the comprehensive documenting, and the definition of the environmental condition and problem based on research assessments. The DMADV approach in monitoring the historic buildings would be also introduced to ensure the preserving quality and process.

6.1 Conservation Levels and Monitoring Process

The research used Zoom Out and Zoom In approaches for monitoring and preserving processing as modes of strategic thinking for advanced leadership. These modes are to ensure the virtues of both perspectives and to create the ability to move across a continuum of perspectives and Top-Down and Bottom-Up Processing, to face the problems of climate change in the coasts, and down to find a solution for local people needs with a whole image perception. Our research approach is shown in Fig. 6.1.

M. Hassan and H. Xie, *Climate Change and Conservation of Coastal Built Heritage*, https://doi.org/10.1007/978-981-13-8672-5_6

Fig. 6.1 Zoom Out and Zoom In approaches and levels for monitoring and preserving processing

Preventative actual actions to protect the natural and cultural resources in the national-level planning, particularly of the flood risk, need high funding could be exceeded the cultural value of the resources themselves. However, the anticipatory national-level planning for adaptation forms the first guide to its potential sustainability. The effectiveness of the national level could be in minimizing the risks and enhancing the future security where there are many constraints on individual adaptation, meanwhile, the efficiency could be in details and small scales where the combined individual actions could contribute to making a significant adjustment.

The sustainable conservation development at the regional level does entail not only compact and smart conservation and land growth but also intelligent treatment, solutions, and design at the down levels. The instantaneous reaction of the locals in the catastrophic circumstances, which is built on their innate and acquired knowledge and at best of advance people preparation for the pre-educated behavior, is fast, cheap, and basic action. On the other hand, the strategic planning has high expensive and slow due to long-term planning and implementation. The conservation procedures, analysis phases, and transition agendas need monitoring, evaluating, and supervising in each step. The conservation team includes a group of creative actors. It consists of locals, people from governments and research institutions, as well as experts, developers, and consultants from different specializations, such as spatial planning, architecture, engineering, conservation, ecology, tourism, and private companies.

Protection and preserving the cultural heritage of climate change should have conservation action levels as shown in Fig. 6.2. The action levels should be hierarchically concatenated from the global action, in term of climate change, rising water level, and the calls of energy efficiency, and going down to the national, local, and municipal level, then microlevel which includes the measures of the site and buildings and their locals. First of all, the global efforts for protecting our planet from climate change and global warming risks must be combined. Global strategies and deterrent measures should be applied to minimize the threats by the great states which have a

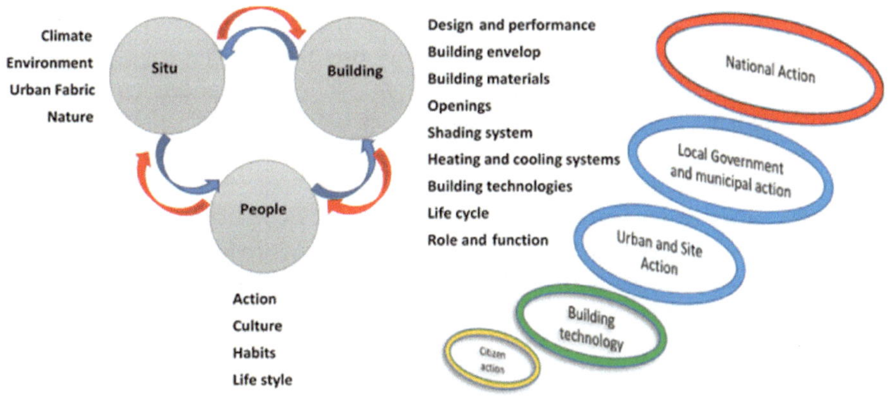

Fig. 6.2 Levels of conservation action and interventions

big negative ecological footprint on our planet and contribute to this problem. The national and regional level concerns the whole climate change risks not only in the built heritage environment and structures but also on the demographic and economic dimensions. In this level, strategic planning, studying the financial burdens, hard sustainable infrastructures, and policies and providing funding for protection strategies, preventive and treatment measures, and actions should be considered, planned, and applied. The knowledge-based economy should be the base and frame for all conservation measures, where science and technology are the base of the sustainable solution, and universities and research institutes should be the leading actor in the conservation process. In the Syrian coastal region, the population increase will result in additional pressure, local ecology impact, and a demand for new ones and other recreational and business functions in the area. Additionally, the investment of oil and gas near the coastal heritage sites should be with a low environmental impact. Building additional facilities in the region territorial water can increase the region area and protect the heritage sites.

At the municipal level, a balance needs to be found between ecological preservation demands and housing needs. Urban and recreational developments could combine with large-scale nature development and can strengthen both environmental and economic values in the coastal areas which already have a long history and a unique location and nature. The developing of the sustainable hard and soft infrastructure, the preparation of the environmental and thermal works of the site, (for instance the site materials and surface designs using green surfaces, highly reflective surfaces, materials with high export capacity and reflective material for pavements), and the enhancing of the residential sector quality would contribute to the resilient city development. The micro- and technical level includes the locals and occupants action, and the measures of improving the quality of the internal environment of the individual historic building, and the building envelope and upgrading its components. This level is a very detailed level. It consists of glazing and shading and façade work, wall insulation shading works, daytime and natural lighting, roof insulation, optical solar devices, optical glare control, exterior paint works in light colors, opening and natural ventilation, ventilation systems, internal ventilation quality, clean materials and chemical contaminants, noise and water control and reduce the air pollution, and carefully choosing the proper cooling and heating works.

The participatory and cooperation among the action levels are helpful and productive, as well as the people power and positive passive behavior could be the significant and fastest and cheapest action in the preservation process. Starting simple steps and small-scale field measurements and experiments where several technologies can be used for innovated conservation will enhance knowledge-based operations and form the solid base for the bigger scales and further steps.

6.2 Regional Conservation Framework of the Threatened Coastal Historic Sites (Macrolevel)

The Zoom Out approach with top-down processing as a stepwise approach for regional conservation and risk management for the coastal heritage sites, which is controlled by higher levels. It is where urban and regional planning becomes a significant key to risk management and conservation. The approach involves making strategies and decisions after looking at a wider scope, and then in the smaller areas wherein more details are added for accuracy. By following several stages of analysis, the results form the basis for the development of future plans for the protection of coastal cities and could be achieved according to the adoption of sustainable development and the proposed defences in the face of SLR problem. The area between the shoreline and defences resulting from the exploitation of the possibilities of the continental shelf features, to urban development area based on clean energy and always work under the blue communities address and ECO-Land conditions. Our research strategy commensurates with the geographic, economic, and political nature of any country and can be generalized and applied in any coastal area around the world after following a series of study and analysis stages, see the flowchart in Fig. 6.3. The proposed sustainable defenses progress and customize with the problem and the proposed sustainable solutions by creating defensive communities of the population and vital projects. Protection strategy for the coastal areas and disaster containment would be at points we identify rather than nature. The approach procedures are as follows:

① Determining the risk and the problem definition based on the data collection and analysis to determine the local climate problem by the people participatory during a distributed survey and based on related literature to cover the lack of information.

② Determining the threatened heritage sites and the preliminary needs and then classifying their values of the threatened heritage sites to assess the conservation priorities.

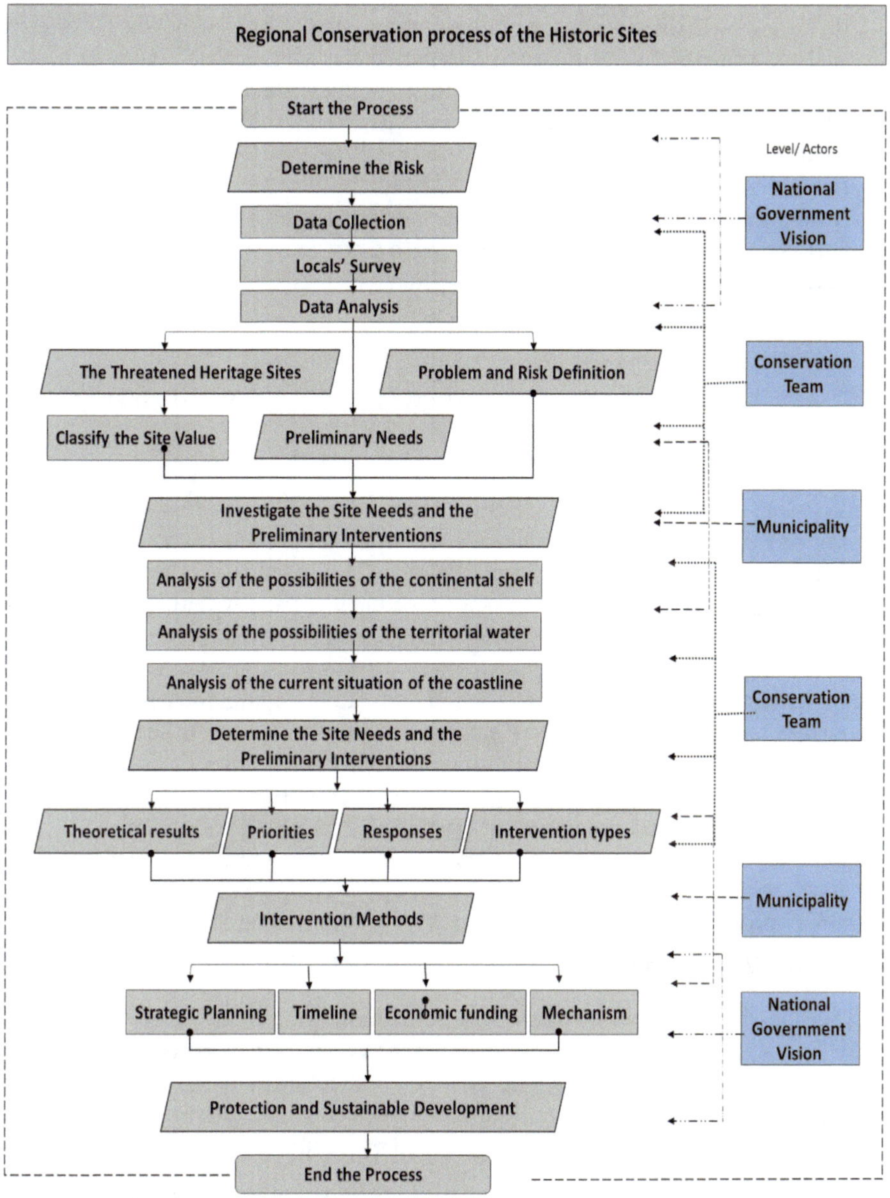

Fig. 6.3 Flow chart of the protection strategy for the coastal areas procedures and intervention levels

③ Investigating the needs and the preliminary interventions by using the descriptive analysis of several steps to determine the current situation and the possibilities of the continental shelf, territorial waters, and the state plan investment.
④ Determining the needs and the preliminary interventions and the theoretical results, the priorities, responses, and intervention types in both wide and small scope.
⑤ Determining the intervention methods, mechanism, timeline, strategic planning, and its economic funding, down to propose a protection strategy and sustainable development to face the problems of climate change on the coasts.

6.3 The Conservation Framework of the Historic Buildings (Microlevel)

The Zoom In approach works on details and uses the bottom-up monitoring and preserving processing upward and controlled from the low levels of the built heritage, historical building microclimate, and occupants' satisfaction. In the low level, the information and decisions are informed by occupants' survey and realistic discussions starting from small scope with all the needed details and incoming data from the environment to form a perception. Then it grows eventually in complexity and completeness to meet local optimization and create the comprehensive vision. The framework of the microlevel aims to enhance the interventions in private spaces with a focus on thermal comfort conditions and livability of the current situation, where the residents can benefit and managing comfort conditions. The key and control strategies, adjusting and preservation approaches, and interventions of the historic vocabularies should be selected based on a sum of data and analysis levels to size the vocabulary situation and identify its problems. The conservation of the historic buildings' procedures, as shown in Fig. 6.4, follows this order:

① **Documenting and Definition**

The documenting step includes data collection, historic literature, and records (previous repair history, previous maintenance logs, old reports, and records), location data (geographical/climatic zones), architectural documenting (drawings, descriptions, etc.), and structural and mechanical documentation. Definition step includes determining the visible problems and the methodology of investigation.

② **Monitoring and Measuring/Determining the value and possibilities**

This step includes the value assessment, comprehensive assessment and diagnosis of the structural condition and building performance based on research, field measurement, and survey (observing the indoor thermal parameters, including air temperature, air humidity, and wind speed in the historic buildings, additionally, the clothing, thermal comfort sensations that determine the comfortable climate conditions). It also includes assessing the buildings environment, envelope possibility and

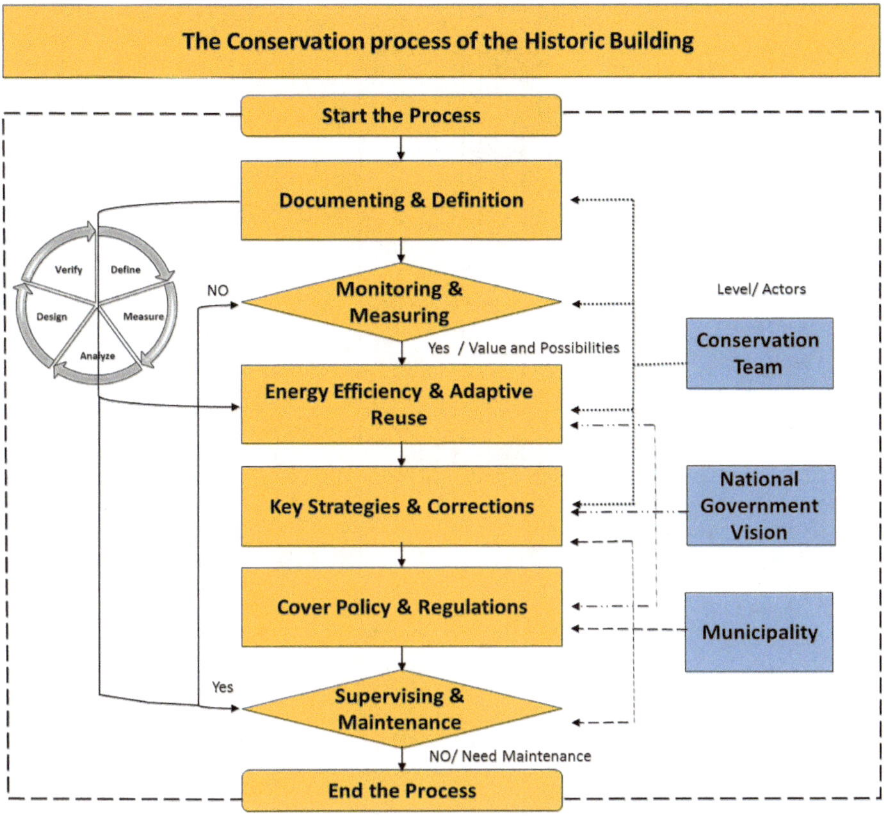

Fig. 6.4 Framework of the historic building conservation (microlevel)

problems, current and predicted risks, and determining what areas need attention, the proper usage, and the corrections and changes.

After the documenting step, two processes (traditional or integrated conservation process) according to the available time and budgets could be followed. The traditional conservation process is linear for a serial collection of separated tasks where the cost decisions driven by initial capital cost or the owner funding. In the traditional method, the building is a collection of separate components and materials, the conservation team generally limited to architects and engineers, and the solutions yield optimized independent systems. The integrated conservation process of DMADV process is cyclical where collaborative, conscious effort to integrate design strategies between all disciplines, and the conservation team includes owner, architect, engineers, construction team, facility management team, users, etc. In the integrated process, consider the building as a whole, life-cycle cost drives the cost decisions, and the solutions yield multiple benefits resulted from the process circle, see Fig. 6.5.

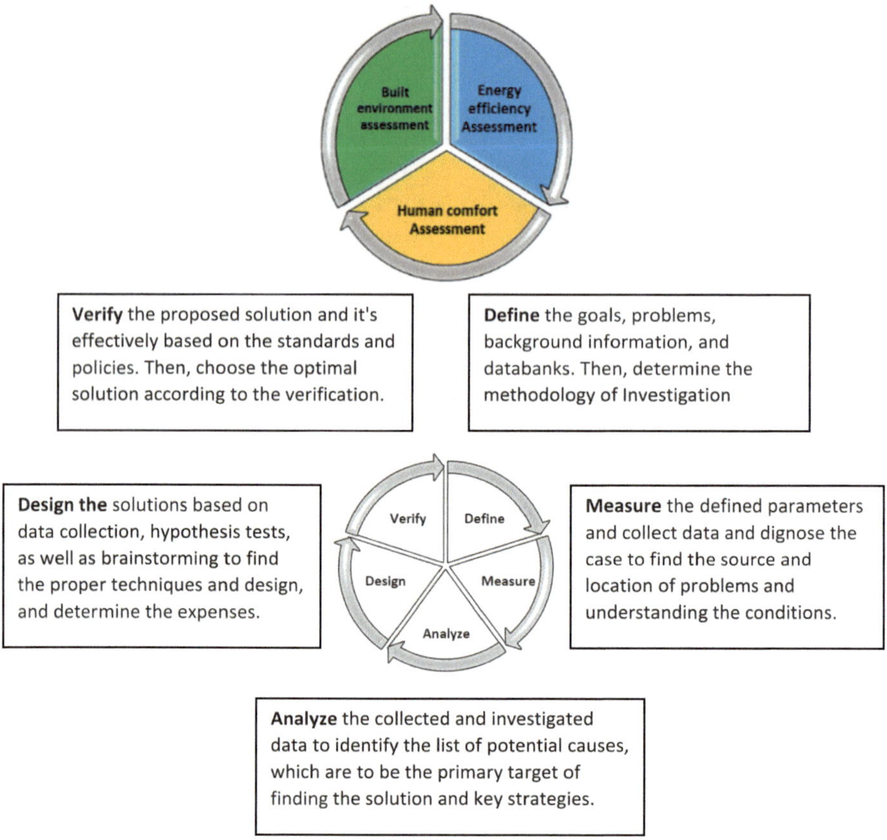

Fig. 6.5 Five conservation phases of the DMADV circle (Define, Measure, Analyze, Design, and Verify)

③ Energy efficiency and Adaptive reuse

This step includes energy efficiency and adaptive-reuse possibility, needs and budgets, retrofitting and proper reuse, interventions to upgrade the structures, cost-benefit analysis of labor and materials, and classification as ideal candidates, middle, and un proper.

④ Key strategies and Corrections

This step includes the design development and correction of the indoor and outdoor spaces, and the implementation mechanism focusing on the following key related aspects: shading, vegetation, wind and ventilation, criteria for materials and colors used for surfaces, and building envelope and components which have a direct effect on thermal comfort. The proposed solutions must use the local materials and production and consider the economic and social capacity of the society to maintain and enjoy these spaces where the owner and users' funding is limited. Furthermore, they should include a cultural and educational task for the locals' starts with policy makers, local inhabitants, and designers.

⑤ **Cover policy and Regulations**

The regulative step includes defining (regionally based) sustainable design codes and taking into account the role of the community in using and maintaining space, codes and guidelines of rehabilitation, structural and mechanical considerations, energy efficiency measures, preserving the historic character and the proposed changes, additions considerations, replacement considerations, supervision, management, and periodic maintenance.

⑥ **Supervising and Maintenance needs**

The supervision and maintenance step includes the municipal supervision role and the community using and maintaining role, and in case the building needs maintenance, the maintenance should be based on the measurement and academic verifying phase.

In general, the adaptive-reuse and conservation team includes different related actors, for instance, history documenter, design professional, restoration contractor, historic conservator, the local preservation office, project manager, structural repair and restoration, strengthening and protection, preservation technology, and owner and buildings' occupants, where the energy efficiency rules, the key strategies and corrections, and the cover policy and regulations follow the national government vision and happen under the municipality supervision and codes.

⑦ **The DMADV approach in monitoring the historic buildings**

Using the DMADV circle (Define, Measure, Analyze, Design, and Verify) could facilitate the conservation process and ensure the preserving quality and solution trust based on data and knowledge.

(1) **Define step** includes the goals, problems, and background information. It consists:

- Databanks/building description and documenting,
- Determining the value,
- Occupants and comfort needs,
- Determine the visible issues,
- Refer the financial aspects,
- Determine the methodology of Investigation.

By the end of this step, all the importance and motivation of this study are considered, and the limitations have been placed (budget, time, and resources).

(2) **Measure the defined goal** to collect data and find the source and location of problems and understanding the conditions. It consists:

- Data collection,
- Diagnostic tables and charts,

- The measured parameters (at least air temperature, relative humidity, and wind speed),
- Occupant's satisfaction survey,
- Calculations increase the trust of the academic study product and the required corrections.

At the end of the measuring step, the current problems based on data which support our approach and form a baseline performance of the process could be defined.

(3) **Analyze step** includes interpreting the collected and investigated data to identify the cause of the problem, which is to be the essential target of the improvement.

- Determine the potential causes to find the solution and main strategies.
- Verify the causes by simulation so that improvement focuses on the deep purpose.
- Determine the problems of the location, climate, building envelope, and windows performance.

By the end of this step, a list of potential causes to find the solution and key strategies would be determined.

(4) **Design step** consists of designing, developing, and implementing proper solutions.

- Determine the required corrections and changes to improve the built environment.
- Generating solutions is based on data collection, hypothesis tests, as well as brainstorming to find the proper techniques and design.
- Determine defects correction costs and other expenses help in making a decision.

By the end of this step, the changes, solutions, and the interventions would be identified.

(5) **Verify step** ensures that the problem will be solved by the proposed solution and determine the solution effectively.

- Verify the proposed strategies and solution and check if the changes integrate to all standards and compliance policies.
- Experimental verification if available is the best way to ensure optimal design.
- The proposed solutions could be evaluated using the simulation and heritage priorities and authenticity. Numerical simulating can improve the solution and reduce the risks.

By the end of this step, the optimal solution would be determined, and the proposed solutions that will solve the problem and lead to improvement would be verified with data.

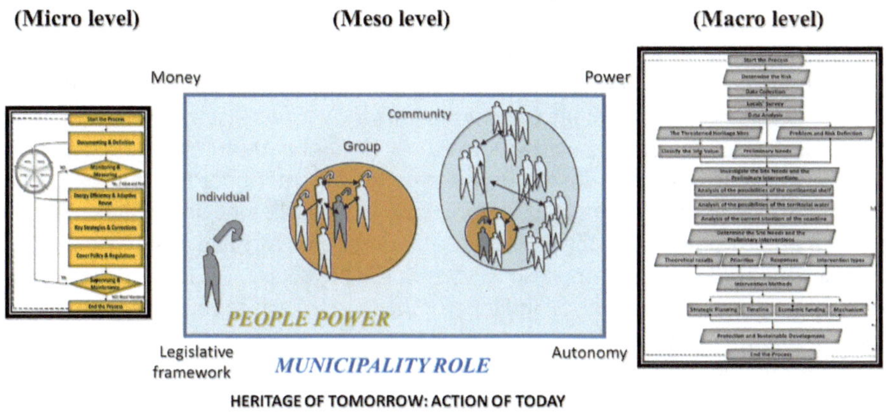

Fig. 6.6 Mesolevel

⑧ How to connect between macro- and microlevels??

The linkage between macro- and microlevels (mesolevel) is based on activating both the locals' power and the municipality role in adjusting and preserving the historic site and buildings, as shown in Fig. 6.6. The proper conservation legislation and measures to accommodate the current usage, buildings types, climatic variations, and the thermal comfort in consideration to the occupants' needs should be established and implemented in the landscape, urban area, or old town, where the municipality has a significant role in changing and implementing the conservation strategies using its authority power points (autonomy, money, power, and legislative framework). To improve our old cities, make them more resilient and dynamics, and make them able to face the challenges, improving their social structures and raising their cooperation to save themselves and their heritage are an essential issue. By making the linkage between the current and future challenges, and by showing the differences between the past efficient experience and the current wrong behavior inside the historic, an actual method to change the locals' daily lifestyle and action can be achieved and helped to learn and bring these lessons in practice. The civilians and professionals from government and other authorities can work together successfully as long they know and accept their role, on different scales and different time schedules.

6.4 Summary

The research is the first study combined between the macro- and micromethods using a twofold approach (Zoom In and Zoom Out) for preserving and monitoring the coastal built heritage sites and buildings of the climate change. It determined the difficulties and needs, studied the impact of heat and humidity on the historic buildings, and proposed a long-term protection strategy of the coasts of climate change at the macrolevel. Thus, it provides a guiding outline for risk management and a reference for the protection of architectural heritage in the coastal heritage sites. Meanwhile, with microclimate research as a start point at the microlevel, the environmental risk and thermal comfort assessments for various historic buildings in the coastal heritage sites could be conducted through field measurement and questionnaires. Our approach provides a guide for collecting and recording the first-hand data of the thermal environment of historic buildings worldwide and in particular for developing countries where the environmental research is rare and no existed thermal standards or guideline. The approach provides a basic dataset for further studies and for establishing future relevant standards for the thermal environment of both coastal historic and new buildings.

In the end, this book presented the minimum set of documentary data elements on historic site and structures need to be collected to have a view before evaluating the climatic risk and doing both preventive and treatment conservation. Our research' results could be useful for future studies' framework and as a baseline for future local standard and measures. Our research also gives a positive human impact on the locals throughout their contributing to determine the site climatic problem and to evaluate the environment quality. By using surveys, people becoming more aware of the importance of their activities and adapting to their environment. Additionally, the evaluations of the climate change problems and thermal comfort by human subjects and evidence are a superior contribution to knowledge, having longer-lasting value to the research community than theoretical evaluations coming out of a theory and simulation model.

Appendix A
Climate Change Survey

Questionnaire No:…

Survey of the Effects of Climate Change on the Tartous Coast in Syria

This survey aims to explore reliable data on the value of Arwad's Cultural Heritage and the effects of climate change. In aspirations to improve the situation of Arwad Island and develop a strategy to protect it from the third immersion. The results of the statistical evaluation, the analysis and conclusions will be published. Please carefully read each question before answer.

M. Hassan and H. Xie, *Climate Change and Conservation of Coastal Built Heritage*, https://doi.org/10.1007/978-981-13-8672-5

Part 1: Personal and Climatic data:

1. **Sex**: □Male □ Female
2. **Age:** □Under 20 years □20-34 years □35-49 years □50-64 years □Above 65 years
3. **Education and training:**

□secondary school □ technical craft skills □ University graduate degree □ Master or higher

□Others, please specify.......................

4. **Is your professional activity related to cultural heritage?** □Yes □No
5. **Specify your professional activity:**

□Materials and restoration □Contractor □Tourism, events, hotel, restaurant

□Site or asset management □Research, university □Service, technology provider

□Public administration, designer, engineer, architect □Others, please specify area of expertise......

6. **Region and place where you live:** ....
7. **Do you visit the island for:?**

□professional reasons □private reasons □Trading □Tourism □work □Others, please specify..........

8. **What is the reason of concerning in Arwad?**

□ I'm from Arwad Island and I live in it. □I'm from Arwad Island, but I don't live in it.

□ I've relatives and Arwad island roots □I'm live in coastal area (determine the place:...........)

□ I'm working there □ I've visited this island □I've never had go to this place, but I concern with it.

□I follow this place news on social media (e.g., Facebook, Twitter.....) □Others, please specify......

9. **Do you currently live or work in the island?** □Yes □No

➢ **Climate data, Site Problems and Locals concern:**

1. **In your opinion, what kind of climate problems affect the island during the summer months?**..
2. **In your opinion, what kind of climate problems affect the island during the winter months?** ....
3. **Could you please provide us a piece of evidence about the climatic problem that the island suffered from?**

   ....

   ....
4. **Do you have any suggestion to the government for:**

- Protection of the island and heritage assets ....
- Oil and gas investment near island ....
- The use of alternative energy on the island ....

5. Do you leave the island for temporary period because of rising sea level in winter?

□Almost always □Often □Sometimes □Seldom □Never

6. If you leave the island for temporary period where do you go?

□Own house in Tartous city □Own house in other city □Relatives house in Tartous city □Relatives house in other city □Abroad □Others, please specify.................................

7. Do you notice rising sea level now more than before? □No □Yes

8. How concerned are you about rising sea level and climate change issue?

□Extremely concerned □Very concerned □Moderately concerned □Slightly concerned □Not at all concerned

9. How concerned are you about the island current situation?

□Extremely concerned □Very concerned □Moderately concerned □Slightly concerned □Not at all concerned

10. Which sources of information do you use to know the weather? (Check all that apply)

□International sources and advertisements □Government weather and national meteorologists □Local people or Family recommendations and experience □Internet and Social networking websites □Others, please specify.......................

11. Are you satisfied with your government works and strategy for protection from rising sea level RSL? □Very satisfied □Satisfied □Neither □Dissatisfied □Very dissatisfied

12. How willing are you to change your lifestyle to reduce the damage that the environment causes to the island and your building?

□Extremely willing □Very willing □Moderately willing □Slightly willing □Not at all willing

13. How likely are you to share in the process of protection the island and its building island starting from your house?

□Extremely likely □Very likely □Moderately likely □Slightly likely □Not at all likely

Part 2: Heritage Assets and Site Investment

1. **What kind of heritage assets you see it should protected?** (Select the 3 most important, prioritize by giving t score 1 to 3, where 1 is most important)

□ Historic buildings □ Defensive heritage (castle, tower, walls, etc) □ Religious assets. □ Civil infrastructures (routes, corniche…). □ Industrial heritage (boats,…, etc).	□ Ruins and buildings from different historic periods (Ancient and prehistoric, Roman, medieval, ... etc.). □ Cultural landscapes. □ Customs and Traditional arts and food. □ Others, please specify………………………

2. **What do you consider negative effects of climate change on the architectural vocabulary of the island?** (Select the 3 most important, prioritize by giving the score 1 to 3, where 1 is most important):

□ Demolition of Phoenician defense wall and sinking the Phoenician monuments of the island because of high waves

□ The degradation of the building facade of the seafront

□Deterring residential use and environmental degradation.

□ Others, please specify…………………

3. **On the basis of your personal preferences, what do you consider negative effects of climate change on the social life of the island?** (Specify the 3 most important):

....

………………………………………………………………………………………………

4. **Could you determine the needs of the cultural heritage sites to be a destination for tourists and to have a positive effect?** (Select the 3 most important, prioritize by giving the score 1 to 3, where 1 is most important):

□ Creates jobs □Attracts tourists and visitors □Nurtures identity of a place □Improves the market for local products □Preserves the urban vocabulary and the history □It is an attractive environment for residents. □ Reduces the need for new buildings by promoting the existing buildings. □ Create new projects and investments. □ Others, ……………………………

5. **How safe would you feel if the oil and gas field investment were built near where you live?**

□Extremely safe □Very safe □Moderately safe □Slightly safe □Not at all safe

6. **Which of the following alternative energy sources do you think will be most important in the next 10 years?**

□ Wind □Solar □Nuclear □Ethanol □Natural gas □Coal □Others, please specify……………………

7. **Should the government provide more money to support alternative energy?**

□ Strongly disagree □Disagree □ Neither agree or disagree □ Agree □ Strongly agree

8. **How often do you use heating and cooling systems in your house during extreme temperatures?**

□Always □ Most of the time □ About half the time □Once in a while □Never

Part 3: Priorities and Government Role:

1. **What actions can the authority take to build a better planning and strategies?**

....

2. **What projects would you like to see in Arwad Island?**

....

3. **Who do you think are the most relevant actors in the sustainable preservation and use of heritage assets?**

(Select the 3 most important, prioritize by giving the score 1 to 3, where 1 is most important):

□ Universities □ Restoration professionals. □ Tourist agencies. □ Local authorities.

□ Private owners. □ Public owners. □ Cultural workers. □ Managers of heritage assets.

□ Local residents. □ Others, please specify…………………

4. **How do you think Defensive Island can foster local development?**
5. **Why is it beneficial to know the economic value of cultural heritage in Arwad Island?**
6. **Do you agree with the Key Priority Areas?**

- Raise awareness and understanding of our heritage and the contribution it can make to everyone's life.
- Strengthen the partnerships and resources that support heritage.
- Ensure that heritage plays a full role in strengthening our communities.
- Ensure that our heritage is conserved for the benefit of future generations.

□ Strongly disagree □Disagree □ Neither agree or disagree □ Agree □ Strongly agree

- **If you disagree, what other Priority Areas do you suggestion?**

....

....

➢ **If you are decision makers, could you answer these questions please?**

1. **Area of making decision and occupation:** ....
2. **What are your reasons for concerning in Arwad Island?**

□Authenticity and Value □Tourism □Economy □Urbanization □just work □ Others, please specify…………………

3. How concerned are you about rising sea level and climate change issue?

□Extremely concerned □Very concerned □Moderately concerned □Slightly concerned □Not at all concerned

4. How concerned are you about the island current situation?

□Extremely concerned □Very concerned □Moderately concerned □Slightly concerned □Not at all concerned.

5. Are there any ideas or projects put to use at your company to deal with climate change and rising sea level problems?

□No at all □Just ideas without applying □Occasionally □some projects □ A lot of projects

6. How effectively were the projects and ideas put to use at your company to deal with climate change and rising sea level problems?

□Extremely effectively □Very effectively □Moderately effectively □Slightly effectively □Not at all effectively

7. How easy was it to get data and financial resources concern the environmental issue for your project?

☐Extremely easy ☐Very easy ☐Moderately easy ☐Slightly easy ☐Not at all easy

8. How much professional expert did your company have or deal with the environmental issue?

□Never □Rarely □Occasionally □A moderate numbers □A great deal

9. How are the national works and research in the environmental field?

☐Extremely effective ☐Very effective ☐Moderately effective ☐Slightly effective ☐Not at all effective.

10. How are the decisions made by the government to support the environmental issue?

☐mandatory ☐not mandatory

11. What actions can the authority take to build a better planning and strategies?

…………………………………………………………………………………………………

…………………………………………………………………………………………………

…………………………………………………………………………………………………

…………………………………………………………………………………………………

Date: …../……/ ……..

Thank you for your support and cooperation and the time you had to devote to this questionnaire.

Maya Hassan

Appendix B
Interview General Questions

In case you live/work in the island, could you answer the rest of questions please?

1. **The building/room location:** .... **Floor of building**: ....

2. **How long you live/work in your present building?**

□Less than 2 years □ 3-10 years □ 11-20 years □ 21-30 years □ 31-40 years

□41-50 years □ 51-60 years □More than 60 years

3. **What is the type of building you live or work in?** □Modern □old building

4. **Building materials:**

□Sandstone □sandstone just as external skin □Industrial stone and Concrete □Wood

□Mixed(sandstone and industrial stone) □ Others, please specify..................

5. **Could you please briefly introduce your previous restoration experience of your building?**

....

....

6. **Does the island local code include the heritage and environmental consideration?**

....

....

7. **What is the satisfaction level of the indoor environment of your building:**

□Very satisfied □Satisfied □Neither □Dissatisfied □Very dissatisfied

8. **Specify the reason of satisfaction or dissatisfaction:**

....

....

M. Hassan and H. Xie, *Climate Change and Conservation of Coastal Built Heritage*, https://doi.org/10.1007/978-981-13-8672-5

9. **Could you please briefly provide us some problems in the indoor environment of your building that you suffered from, and what kind of visible damage your building has?**

....

....

10. **Do you have any suggestion to enhance your building condition?**

....

....

Appendix C
Thermal Indoor Environment Survey

Thermal indoor environment survey (Summer Season) **Survey Number:**...

The purpose of this survey is to evaluate the thermal conditions in the historic buildings in Tartous, through the information collection from the occupants of this building. Please participate in this survey by completing this questionnaire sheets.

Part 1: Basic data:	Occupant Location in Area:
1. Occupant's Name: **2. Date:** - -**2016. 3. Time:**□	(Place an "X" in the approximate place where you live/work.)
4.Age: **5.Sex :** □M □F **6.Height(cm):** **7.Weight(kg):**	Plan changes based on the surveyed case study site
8. Education background: **9. Occupation:**	
10. Is your professional activity related to cultural heritage? □Yes □No	
11.Building location: **12. Floor of building:**	
13.How long did you live/work in Tartous old city?..	**Outdoor Air Temperature(°C):**
14. How long you live/work in your present house/building?...............................	
15.What is the type of building you live/work in?	**Indoor Air Temperature (°C):**
□Modern□Old □Traditional □Ancient □National heritage	**Outdoor Air Speed(m/s):**
16. How old is your house/building?...	**Indoor Air Speed(m/s):**
17.How many rooms does it have?..	**Relative Humidity:**
18. Total Number of Occupants (How many persons do your family/ room have)?	
Part 2: Building envelope and construction:	

M. Hassan and H. Xie, *Climate Change and Conservation of Coastal Built Heritage*, https://doi.org/10.1007/978-981-13-8672-5

1. Building envelope type:

□Sandstone□sandstone just as external skin□Industrial Stoneand Concrete□Wood□Mixed(sandstone and industrial stone) □Others, please specify……

2.Ancient Heating and Cooling System:

□Fireplace,□Wind catcher,□Others please specify…

3.Energy upgrades: □Natural ventilation □No energy upgrades □ Electric cooling devices □Air condition □Manual devices □Others please specify…………………

4.Building orientation:

□South □North □East □West □Two or more directions, please specify ………………

5. Is Building with□Two courtyards □One Courtyard □No courtyard

6. Number of external facades:

□No external facades□One □Two□Three□Four □Separate block(no closed facade)

7. Which side of your house doesn't have a window? □North, □South, □East,□West

8. What is the roof type?□Vault-dome construction □With dooms □Flat roof□Sloping roof □Others, please specify …………………………………………..

9. **What is your roofing material**? □Sandstone □Concrete, □Aluminum plates, □wood, □Hardboard ceiling, □Others, please specify……………………………………

10. External walls type:□Massive Sandstone of double-layer walls□Sandstone of one-layer walls□Artificial Stone □Wood□mud □Others, please specify…………………

15. Please indicate the style of the window in your building:

□Casement window.

□Louver/Shutters, □Aluminum sliding window □Other, please specify ……………………….

16.WindowsMaterials:

□glass with wooden frame □glass with aluminum frame □glass with iron frame □Other, please specify….

17. Shutters materials:

□no shutters □Louver □wooden shutters □just iron bars □Other, please specify …………

18. If there are shutters: what is the shutter design of the windows:

□wide open □just open with angle of 45 away from the wall □always closed□Others, please specify……**…**

11. Are External walls?□Concrete plastered □lime plastered, □with ceramic layer □No insulation, □Paint (**What is the color of the painting?……)**□Others, please specify……

12. Are Internal walls?□Concrete plastered □lime plastered, □with ceramic layer □No insulation, Paint (What is the color of the painting?……)□Others, please specify……

13. How much is the estimated thickness of the External walls?...........cm, **and of the Internal walls**?...........cm

14. How many <u>openings</u> do you have in your<u>house/building</u> in total?

Windows: □No windows □One □Two □Three □Others, please specify …………

Doors: □One □Two □Others, please specify ………………..

19. What method of sun shading do you use?

□Curtains□Veranda/Balcony □Others, Please, specify the reason for using that shading ………………

Part 3: Thermal Questionnaires

1. How would you describe indoor humidity in your Building?

□Much too dry, □Too dry, □Slightly dry,□Just right, □Slightly humid, □Too humid, □Much too humid

2. General Thermal Comfort: How do you feel about the temperature at this moment?

□Cold □Cool □Slightly cool □Neutral □Slightly warm □Warm □Hot

3. How do you feel about the possibility of controlling thermal comfort? □Satisfied □Not satisfied

4. Can you open/close the windows? □No □Yes □There aren't any window

5. How often do you do this(open or close the windows)?□Never □Rarely □Sometimes □Frequently □Always

6. Can you open/close the external doors? □No □Yes □there aren't any doors

7. How often do you do this(open or close the outerdoors)? □Never □Rarely □Sometimes □Frequently □Always

8. Can you open/close the inner doors? □No □Yes □there is no door

9. How often do you do this(open or close the inner doors)? □Never □Rarely □Sometimes □Frequently □Always

10. Can you control curtains/rolling shutters? □No □Yes □There is none of them

11. How often do you do this? □Never □Rarely □Sometimes □Frequently □Always

12. Can you regulate the thermostat? □No □Yes □There is no thermostat

13. How often do you do this? □Never □Rarely □Sometimes □Frequently □Always

14. Can you regulate fan coils? □No □Yes □there aren't any fan coils

15. How often do you do this? □Never □Rarely □Sometimes □Frequently □Always

16. what do you think about energy expenditure in your building?

□Very satisfied □satisfied □indifferent □dissatisfied □Very dissatisfied

17. Which types of Energy consume you more in your building?

□Wood □Charcoal □Sardust of chip □Paraffin □LPG □Electricity

18.What would you like to work saving energy in your home?

□ use space heating/cooling appliances as little as possible □use illumination appliances as little as possible

□use energy efficiency appliance □reduce standby power consumption □no energy saving actions

19.What does "Thermal Comfort Condition" mean to you?

....

....

....

20. General Environment Comments

....

....

....

....

....

Thank you for your support and cooperation

Maya Hassan

3.How do you find it?

□Acceptable □Slightly acceptable □Not acceptable □Slightly not acceptable

4.What is your thermal sensation?□Comfort □Light annoyance □Annoyance □Heavy annoyance

5.How would you like to feel?

□Much too cool □Too cool□A little bit cool□No change □A little bit warm □Too warm □Much too warm

6.On the basis of your personal preferences, how would you consider the room/building temperature?

□Acceptable □Not acceptable

7.How do you consider this room/building?

□Perfectly tolerable □Slightly hard to tolerate □Hard to tolerate □Very hard to tolerate□intolerable

8. If you are dissatisfied with the temperature in your house, which of the following contribute to your dissatisfaction:

□Always too hot, □Often too hot, □Occasionally toohot, □Occasionally too cold, □Often too cold, □Always too cold

9. When is this most often a discomfort problem? □Morning (before 11am), □Mid-day (11am-2pm),□Afternoon (2pm-5pm), □Evening (after 5pm), □No particular time, □Other,(specify)…………..

10.How do you feel about the air flowat this moment?

□Much too still, □Too still,□Slightly still, □Just right, □Slightly breezy, □Too breezy, □Much too breezy

11. Would you like to have an air □Smaller than now □Exactly how it's now □Greater than now

12. Occupant's Clothing: Please provide us the clothing (e.g. Underwear, woman clothes, trousers, accessories, shirt, pullover, etc.) that you are currently wearing as you fill out this sheet. If you are wearing articles of clothing not listed, please enter them into the space provided below:

□Long Sleeve Shirt/ Blouse □Short Sleeve Shirt/ Blouse □Trousers/Long Skirt □Dress □Shorts □Socks □Shoes □Sandals□T-Shirt □Undershirt□Vest□Sweater□Other (specify)…………………………………………….

13. Occupant Activity Level: What was your occupation during last hour?

Occupation	Last 10 minutes	10-20 minutes ago	20-30 minutes ago	30-60 minutes ago
Doing office job	□	□	□	□
Walking in slope	□	□	□	□
Walking in flat land	□	□	□	□
Walking in ascent	□	□	□	□
Resting	□	□	□	□
Running	□	□	□	□
Seating	□	□	□	□
Taking a bath	□	□	□	□
Driving	□	□	□	□
Reading	□	□	□	□
Eating	□	□	□	□
Washing dishes	□	□	□	□
Cooking	□	□	□	□
Doing something else	□	□	□	□
Other (specify)	……	………….	…………	…………

Part 4: Personal microclimatic control and Energy consumption:

1. Which of the following do you personally adjust or control in your house? You can tickmorethan one option:

□Open internal door □Open window □Go outside □Air condition on □Fan on □Portable Fan □Other (specify) ……………………….

2. How can you define the level of control of microclimatic conditions?

□No control □Light control □Medium control □High control □Total control